1 Drivers of change

2 Consequences of change

DK

THE SCIENCE OF OUR CHANGING PLANET

Editor Cathy Meeus
Art Editor Duncan Turner
Managing Editor Angeles Gavira Guerrero
Managing Art Editor Michael Duffy
Production Controller Gillian Reid
Jacket Designer Akiko Kato
Jacket Design Development Manager Sophia MTT
Art Director Karen Self
Design Director Phil Ormerod
Associate Publishing Director Liz Wheeler
Publishing Director Jonathan Metcalf

First published in Great Britain in 2021 by
Dorling Kindersley Limited
DK, One Embassy Gardens, 8 Viaduct Gardens,
London, SW11 7BW

The authorised representative in the EEA is Dorling Kindersley
Verlag GmbH. Arnulfstr. 124, 80636 Munich, Germany

Previously published in 2016 as *What's Really Happening to Our Planet* and 2019 as *How We're F***ing Up Our Planet*

A Penguin Random House Company
10 9 8 7 6 5 4 3 2 1
001-324966-Nov/2021

A CIP catalogue record for this book
is available from the British Library.
ISBN: 978-0-2415-1513-6

Printed and bound in UAE

For the curious
www.dk.com

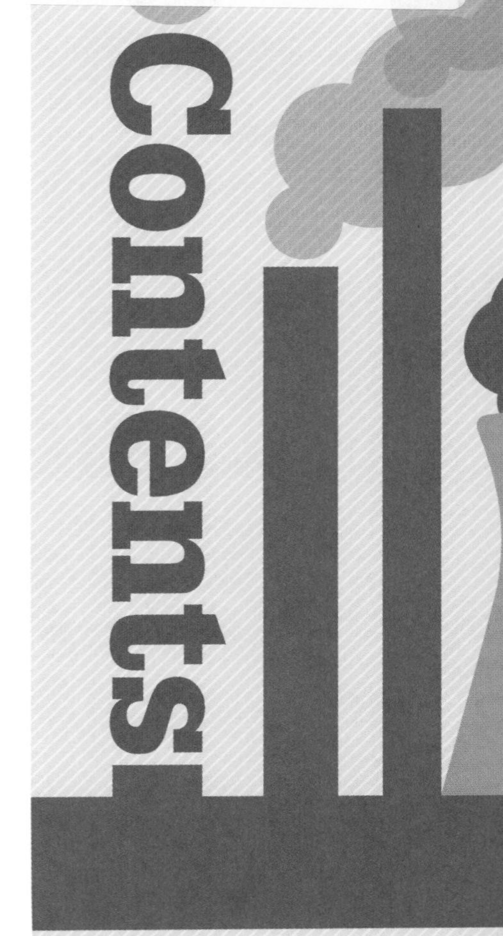

3 Bending the curves

About the author

Environmentalist Tony Juniper, CBE, is Chair of the official nature conservation agency Natural England and a Fellow of the University of Cambridge Institute for Sustainability Leadership. He is a former Director of Friends of the Earth, Executive Director of WWF UK, President of the Royal Society of Wildlife Trusts, and environment advisor to HRH the Prince of Wales. He has also advised many international companies.

CLARENCE HOUSE

Our world is changing faster than at any time in human history, as a result of our own actions. To support a rapidly and unsustainably growing population and develop our economies we are consuming ever-greater quantities of natural resources and making physical changes to the Earth that would have been unthinkable even a few generations ago. There have, of course, been some positive outcomes, including a reduction in overall levels of poverty. But, of course, at the same time, we are seeing many things that give grave cause for concern, including the loss and degradation of natural environments and a changing climate.

We only have one planet to sustain our existence, so it couldn't be more important that we all have good information about the real consequences of human impacts on our life-support system. Only by understanding what is actually happening can we make good decisions, both personally and collectively, about what needs to be done to ensure that we can survive and flourish.

We know that human activity is, however inadvertently, placing unprecedented pressures on natural systems and causing a daunting list of environmental and social problems. Increasing demand for food, energy and water is in many places leading to deforestation, damage to marine environments, pollution, desertification, and the loss of wild species on a massive scale.

I think many people know, in their hearts, that this cannot be right, or sustainable, but the facts that lie behind these and a plethora of other important trends that affect our existence are hard to find. Debates between entrenched vested interests generate more heat than light and discourage innocent inquiry.

In truth, a huge mass of data and insight is available, indeed far more than ever before. It is collected by scientists and specialist agencies and regularly updated, but it tends to be either hidden in technical reports or presented with jargon, acronyms and seemingly disconnected statistics that make little sense to many of us.

I believe we all need to see and understand this information. That includes young people still at school, executives running companies and even experts in particular fields who sometimes don't have time to see summaries of findings from their counterparts working in other fields. We also need to understand the connections between seemingly disparate trends, such as the impact of deforestation on rainfall and the consequences of accumulated plastics in the oceans.

That is why I believe 'The science of our changing planet' is such an important and timely book, bringing together as it does so many rich and authoritative information sources and presenting them in a way that everyone can readily understand. Tony Juniper is an excellent communicator who really knows his subject and presents the information he has gathered in a refreshingly straightforward way. As well as illuminating the problems, he also reveals some of the emerging solutions that need to be adopted while there is still time, including the move to circular economies, in which nothing is wasted.

I do hope that this book finds a broad readership. It contains information that is vital for shaping a positive future for humanity and the rest of life on Earth. Information is, as they say, power, and this easy to read volume will empower in ways that I hope will inspire us to muster the collective will to act – before it is all too late and not once we are faced by escalating catastrophes on all sides.

Charles

Introduction

During recent decades, the pace, scale, and scope of forces shaping planet Earth have rapidly increased. Population and economic growth, coupled with rising demand for resources and the environmental impacts caused by these trends, are rapidly changing the face of our world. If we are to secure our future, vitally important questions must be urgently addressed.

Understanding the scale and scope of the changes shaping our future, and the connections between them is vital to enable us to make sense of our modern world. The unprecedented transformations unfolding around us touch all areas of society, from business and finance to politics and economics, and from science and technology to behaviour and culture.

Drivers of change

Seismic shifts are shaping our future. Rapidly rising population is a key factor. In 1950 there were 2.5 billion people on Earth, today it is more than three times that number, and by 2050 the global population is expected to exceed 9 billion. The impact of people on our planet is, however, not only a result of how many of us there are, but is also affected by our collective standard of living. This is why the rapid expansion of the global economy seen during recent decades is another fundamental driver, enabling more people to enjoy the comforts and benefits that come with increased income and consumption.

Economic growth and rising living standards have in part been fuelled by rapid urbanization, and the progressive shift of people from rural areas to towns and cities. Over the past 50 years the process that

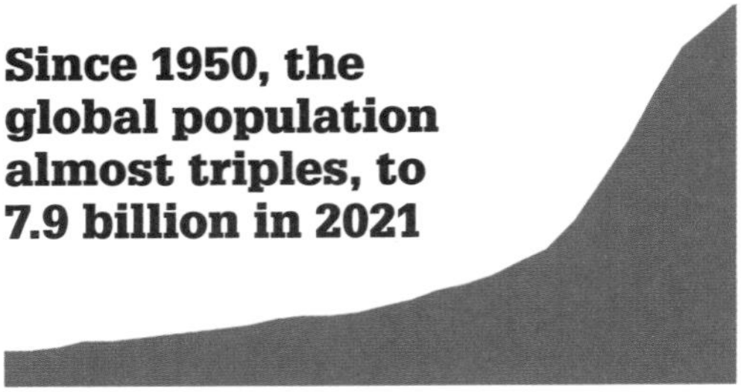

Since 1950, the global population almost triples, to 7.9 billion in 2021

POPULATION EXPLOSION

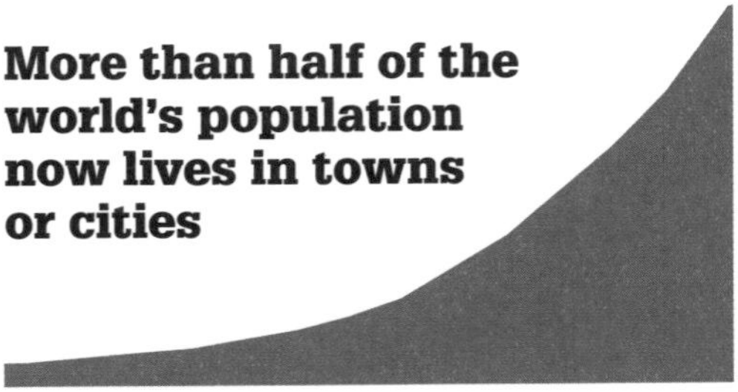

More than half of the world's population now lives in towns or cities

ESCALATING URBANIZATION

began during the 18th century with the Industrial Revolution in England has spread worldwide. In 2007, and for the first time in human history, more than half the people on Earth lived in urban environments. By 2050, the proportion of those living in towns and cities will be approaching two-thirds. City dwellers tend to consume more resources than people in rural areas, using more energy and materials, and generating more waste. Population growth, economic development, and urbanization are key drivers of change that have converged to rapidly increase the demand for a wide range of essential resources, including energy, fresh water, food, wood, and minerals.

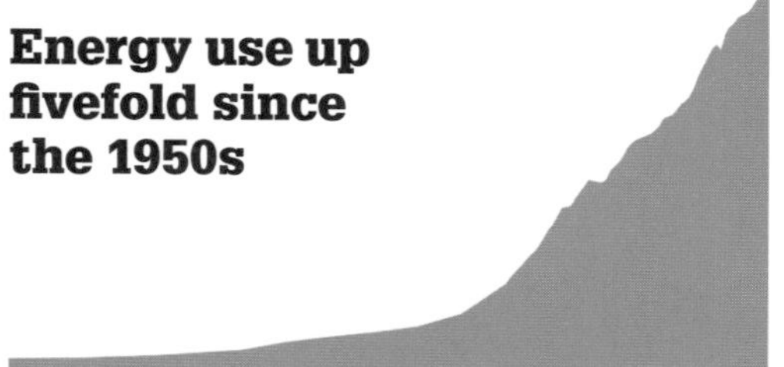

GROWTH IN FOSSIL FUEL USE

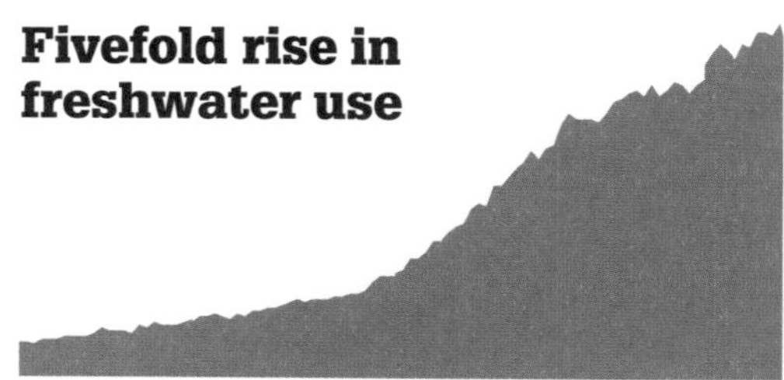

RISING FRESHWATER USE

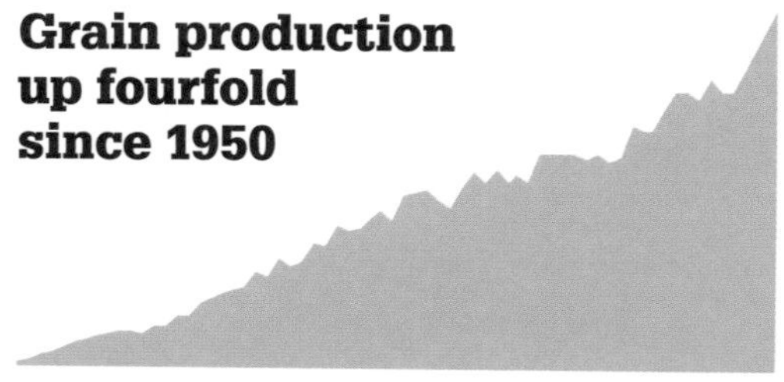

EXPANDING FOOD NEEDS

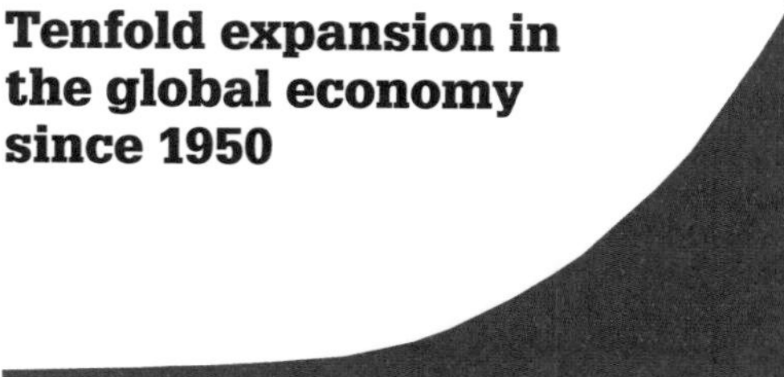

Tenfold expansion in the global economy since 1950

RAPID ECONOMIC GROWTH

Progress ...

Despite concerns about our ability to increase the supply of resources to keep pace with demand, we have so far been broadly successful and over time most social indicators have improved. For example, billions of people have safer water supplies and access to electricity, the number of literate people has increased while the number living in abject poverty has gone down. Various health indicators, such as those relating to child mortality and (notwithstanding the impact of COVID-19) contagious diseases, have improved. We are also more globally connected, with billions of people enjoying access to the internet and social media, and are able to obtain consumer goods traded through supply chains that span the planet.

...and problems

Alongside these measures of progress are a number of less positive consequences. Earth's atmosphere now has a higher concentration of greenhouse gases than at any time for at least 800,000 years. This is already causing climate change, leading to more extreme conditions, increased economic costs, and major humanitarian impacts. The combustion of fossil fuels and forest fires are major factors not only driving climate change, but also resulting in air pollution that kills millions of people every year.

In addition, the depletion of a range of resources essential for human well-being is also leading towards economic and social strains. Freshwater and marine fish stocks are experiencing greater pressures. Soil damage is a worldwide problem, as are deforestation and the decline of species diversity.

The scale of ecosystem degradation means that a mass extinction of animals and plants is gathering momentum. This could soon lead to the greatest loss of species since the dinosaurs were wiped out 65 million years ago. All these environmental changes and many more will increasingly impact on economic growth and development, ultimately threatening to reverse social gains.

Saving the planet

Despite many efforts to find solutions, the process of change is going too slowly. This is because many assume that the environment must be sacrificed to continue to achieve social progress and economic development. And there is resistance from vested

TAKING FISH FROM THE SEA

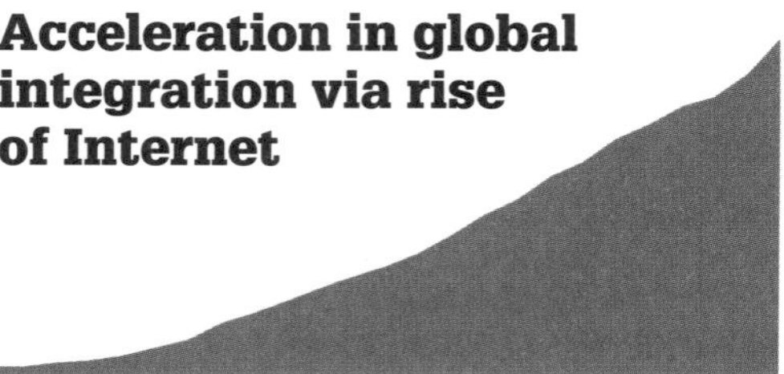

Acceleration in global integration via rise of Internet

GROWTH IN GLOBALIZATION

interests, short-term political choices, and corruption. The need to overcome these barriers becomes more pressing each day.

Achieving change depends on good information and equipping everyone with the facts. That is what this book is all about. In the following pages, I draw on the vast body of data collected by specialist bodies and scientific agencies to paint a picture of our world in the first quarter of the 21st century. Understanding the full range of trends in play is a vital prerequisite for shaping the critical choices we must all make, whether as individuals, companies or governments. I hope readers will feel both inspired and empowered to act in whatever ways they can.

DR TONY JUNIPER, CBE

Tenfold rise in consumption of natural resources

RISING USE OF RESOURCES

Record concentrations of greenhouse gases in the atmosphere

ESCALATING CARBON DIOXIDE EMISSIONS

Human consumption of Earth's renewable productivity doubles

RISING LAND USE BY HUMANS

Mass extinction of animals and plants gathers momentum

SPECIES DECLINE

"The **grand challenges of our age** such as climate change, and the ever-increasing appetite of our planet's **rapidly expanding population** for clean water and energy, require **scientific and engineering solutions** as well as political ones."

PROFESSOR BRIAN COX, BRITISH PHYSICIST AND BROADCASTER

 **The population explosion**

 **Economic expansion**

 **City planet**

 **Fuel for growth**

 **Escalating appetite**

 **Thirsty world**

 **Consuming passions**

1 DRIVERS OF CHANGE

Rapid change is being driven by a series of powerful and interconnected trends. Working together they are transforming humankind's impacts on the natural systems that sustain life.

The population explosion

Of all the trends shaping our changing world, the rapid increase in the human population is perhaps the most fundamental. More people create a greater demand for food, energy, water, and other resources, driving pressures on the natural environment and atmosphere. Although the rate of increase is now slowing down, human numbers rose massively during the 20th century. Our population continues to increase at a rate of over 200,000 per day, or about 80 million a year – annually adding the equivalent of the population of Germany.

Expanding planet

Modern population growth began around 1750, with improved food production and distribution, which lowered mortality rates during the 18th century. The 19th century introduced improved sanitation and other developments that contributed to foster better public health, and during the 20th century the growth rate accelerated at an unprecedented level. It is expected that by 2023, there will be eight billion of us on planet Earth – and by 2050, over nine billion.

The Great Acceleration begins

For thousands of years Earth's human population remained very low and sustainable. This situation changed dramatically, as shown by the massive rise in numbers, from the mid-18th century.

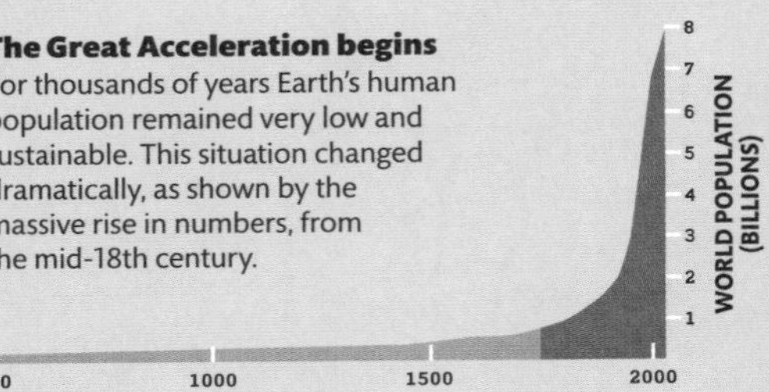

"Population growth is straining the world's resources to **breaking point."**

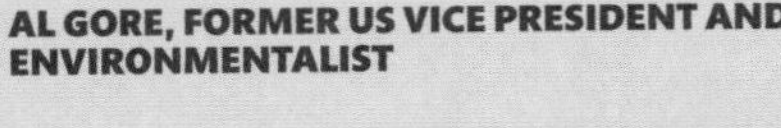

AL GORE, FORMER US VICE PRESIDENT AND ENVIRONMENTALIST

1798
The smallpox vaccine (the world's first effective vaccine) is introduced by Edward Jenner

EARLY 1800s
World population hits 1 billion for the first time

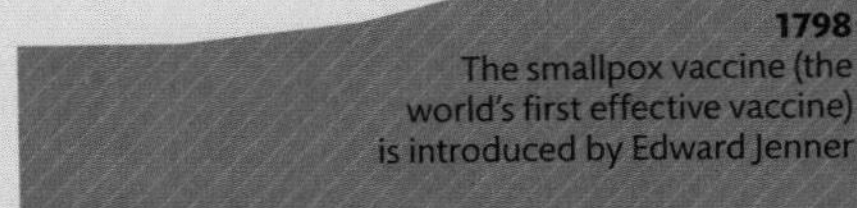

AN INCREASINGLY CROWDED WORLD

In the early years of the 19th century, the world's total population passed one billion. In 1959, the population crossed the three billion mark, and 15 years later, it reached four billion. By 1987, there were five billion people on the planet, six billion in 1999, and in 2011 the total had reached seven billion. By 2020, just five countries were home to more than 3.6 billion people – nearly half of the global total, and three times the population of Earth in the 19th century.

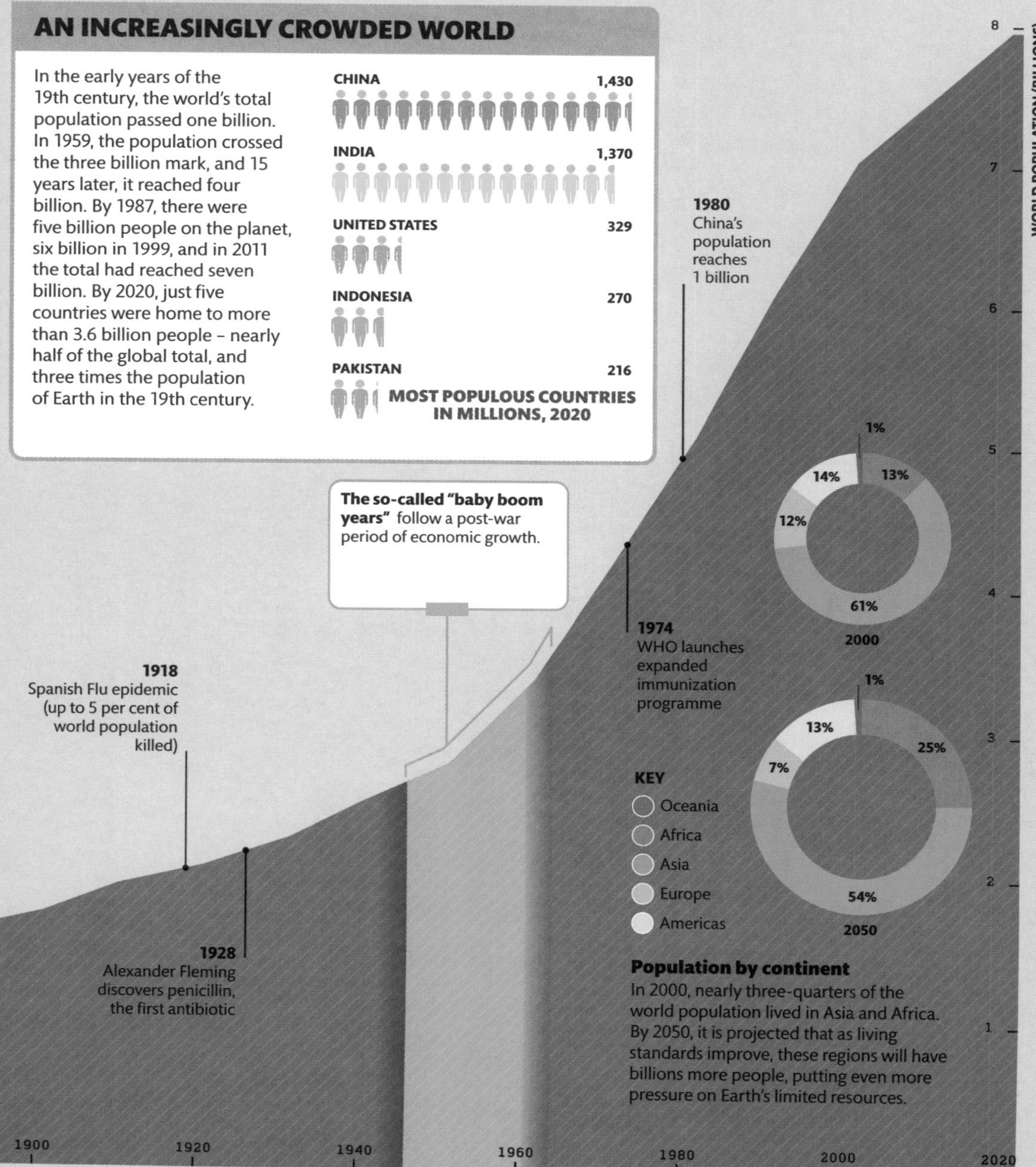

Population by continent

In 2000, nearly three-quarters of the world population lived in Asia and Africa. By 2050, it is projected that as living standards improve, these regions will have billions more people, putting even more pressure on Earth's limited resources.

Population shift

From 1800, population grew in all regions. It began to slow down in richer countries during the 1950s and 60s, as wealth, health, and education drove down birth rates, but growth continued in developing countries.

High birth rates, improvements in medical care, and the influence and movements of migratory workers all contribute to high population growth rates around the world. In the past five years, the biggest population shift has taken place in the Middle East, where the promise of jobs as well as conflicts in neighbouring countries has resulted in the populations of Oman and Qatar rising in excess of 6 per cent a year. While 6 per cent may not sound impressive, at this rate, the population of these two countries will double in 12 years.

USA
0.7%
Current growth rate adds 2.3 million people each year, roughly the population of Houston

Earth's changing profile

Populations in many of the most developed countries are now either stable or growing, primarily because of immigration. The highest percentage growth rates are currently mainly in Africa, which is why the number of people living in that continent is set to more than triple from about 1.2 billion today to more than four billion by 2100. In 2050 roughly 90 per cent of the world population is expected to reside in countries currently regarded as developing (up from about 80 per cent today).

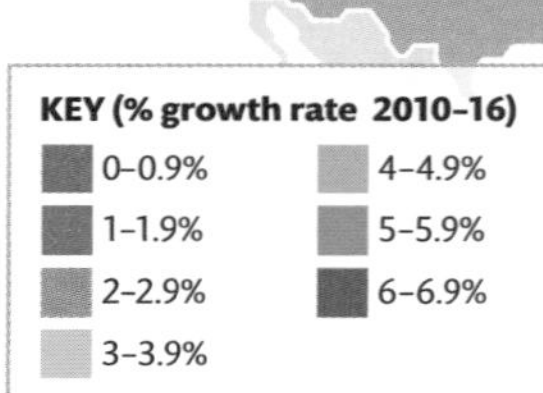

BRAZIL
0.9%
Brazil's birth rate has been falling steadily since the 1960s, reducing its rate of population expansion

WHO LIVES WHERE, PAST AND FUTURE

In 1950, more than 20 per cent of the world's people lived in Europe. By the end of this century that proportion will have shrunk to around 6 per cent. A bigger, opposite scenario is expected in Africa, which by 2100 could be home to almost 40 per cent of humankind. As was once the case in current developed countries, falling death rates will be the major factor driving its population growth.

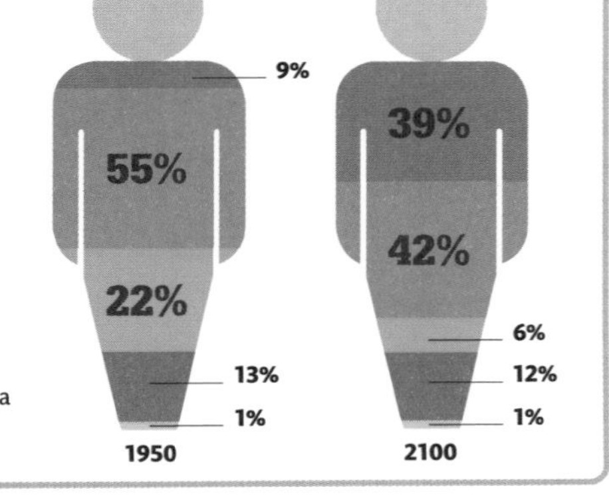

KEY

Percentage (%) of world population

Europe
Americas
Oceania

UK
0.8%
Growth rate equals an extra half a million people each year, roughly the size of Edinburgh

OMAN
6.2%
Oman currently has the highest population growth rate in the world

QATAR
6.1%
The economy attracts rich Westerners and migrant workers from the East, boosting population

KUWAIT
5%
Seventy per cent of the population are expatriates, mainly working in oil and construction

NIGER
3.8%
A fertility rate of more than seven births per woman sustains high population growth

THE GAMBIA
3.1%
At current rates the population will double in about 25 years

BURUNDI
3%
Growth rate is outstripping economic growth and food supply

UGANDA
3.3%
Population to hit 130 million in 2050, from 45.7 million in 2020

UAE
4%
From a peak of 15.2% in 2007, population growth in Dubai is now falling

Centre of the world

In 2015 more than half the world's population lived inside this circle. China and India were Earth's most populated countries, with about 1.4 billion and 1.3 billion, respectively. There were more than 260 million people in Indonesia, over 90 million in Vietnam, and nearly 70 million in Thailand.

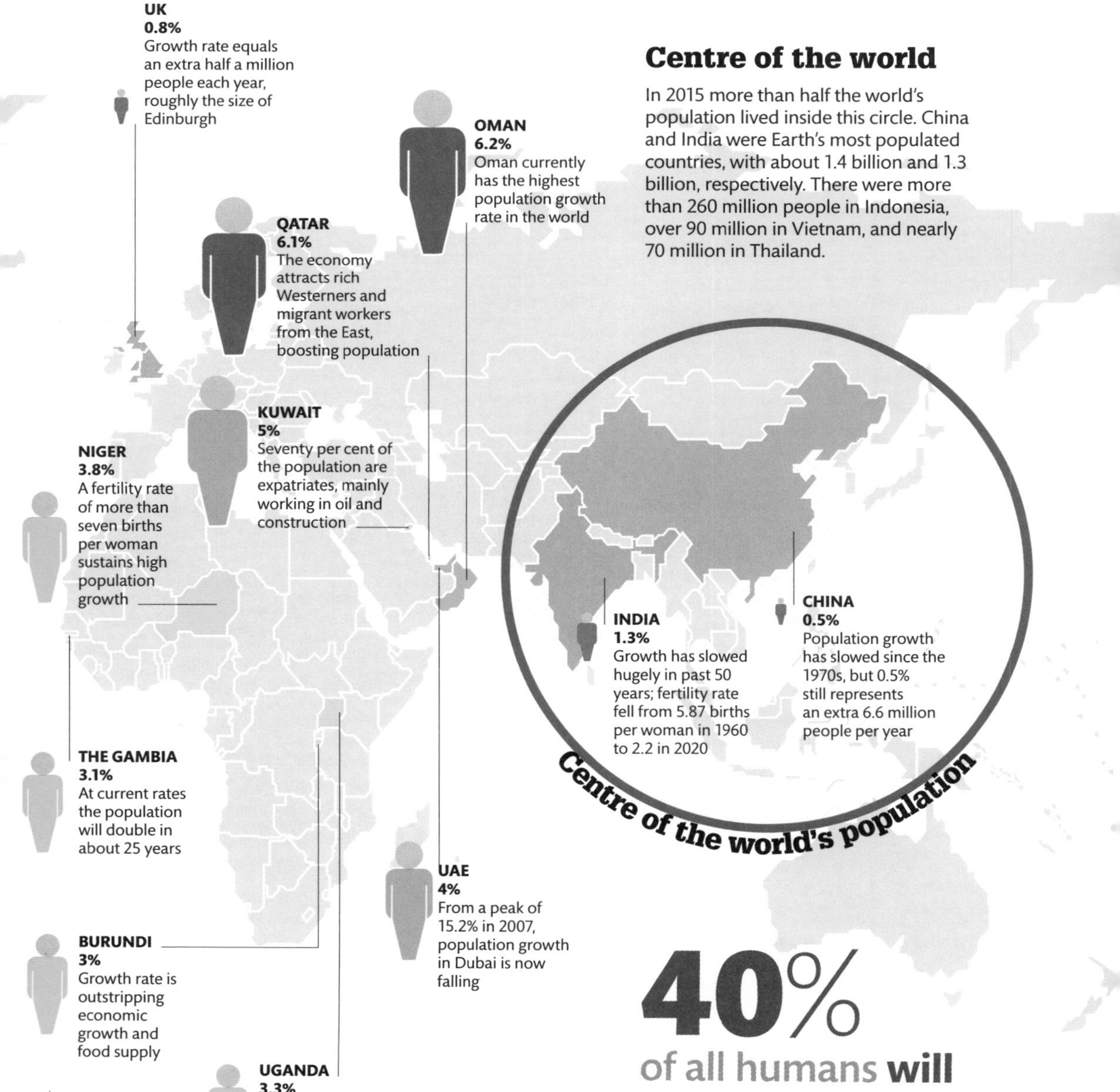

INDIA
1.3%
Growth has slowed hugely in past 50 years; fertility rate fell from 5.87 births per woman in 1960 to 2.2 in 2020

CHINA
0.5%
Population growth has slowed since the 1970s, but 0.5% still represents an extra 6.6 million people per year

Centre of the world's population

40%
of all humans **will be African by the end of the 21st century**

Living longer

Since the beginning of recorded history, young children have outnumbered those reaching old age – at least until very recently. Today, there are more people on the planet aged 65 and older than those aged five and under.

As both the average length of life and the global proportion of older people have increased, a situation without precedent has emerged, raising many important questions. For example, will the aging trend be accompanied by longer periods of good health in old age? Will there be new opportunities for different roles for older people in society? How will societies cope with a much higher proportion of retired people, many of whom will not be paying income tax?

Driven by falling fertility rates and a remarkable increase in life expectancy, population aging will continue to accelerate. Where today's employed population falls typically between the ages of 20 and 65, in future a higher proportion of healthy older people will remain employed, vying with younger workers trying to find jobs.

SEE ALSO...

- **Slowing the rise** pp22–23
- **Better lives for many** pp96–97
- **Healthier world** pp102–103

Life expectancy at birth

Increased life expectancy in the past 100 years reflects a shift in the principal causes of death. In the early 20th century the main causes of mortality were caused by infectious and parasitic diseases. Improved public health, nutrition, and medical breakthroughs such as antibiotics and vaccines have since transformed the situation. Today, people are far more likely to die from non-communicable illnesses such as cancer and heart disease.

KEY
Life expectancy at birth (years)

- World average
- North America
- Latin America and the Caribbean
- Europe
- Oceania
- Asia
- Africa

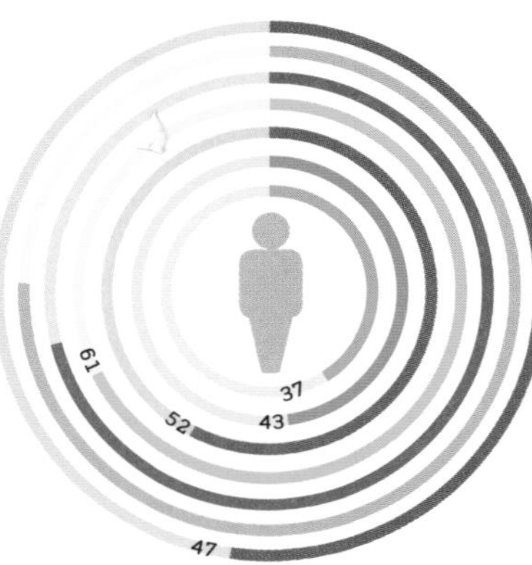

1950–55
North America and Europe exceeded the global average longevity of 47 years by the longest margins. War, disease, and malnutrition all played their parts in shortening lives.

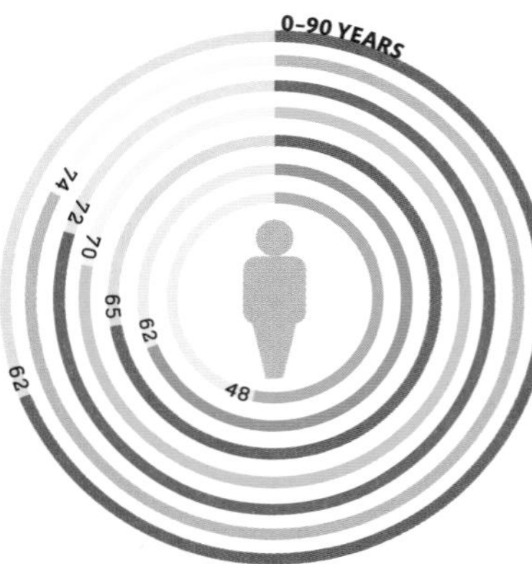

1980–85
Increasingly affluent lifestyles in developed countries, and improved food security and better access to medicines elsewhere, increased average lifespans throughout most regions.

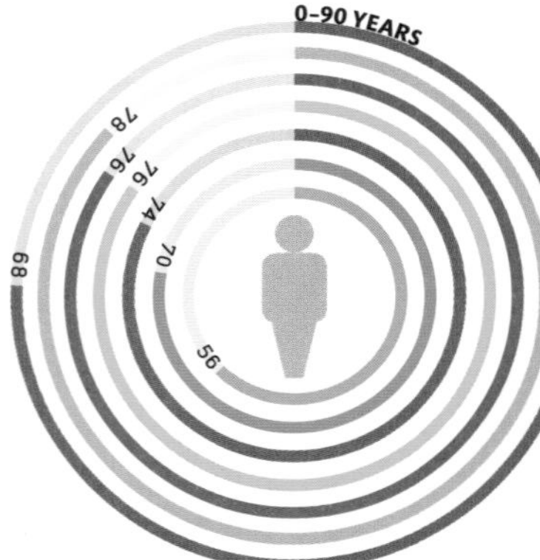

2005–10
Economic growth, better nutrition, and disease control achieved rising longevity worldwide. Africa still has the shortest average lifespan, as many countries remain affected by HIV/AIDS and other diseases.

The world's population in pyramids

The shape of the global population age profile is changing rapidly. The rising proportion of people aged over 60 has caused the pyramid to become not only taller than in earlier decades but also wider at the top. Compared with the situation in 2000, the proportion of people aged 60 years or over is expected to more than double by 2050 to about 21 per cent of the world total. By 2100, that proportion is expected to be about triple.

By 2047 people aged over 60 will outnumber children

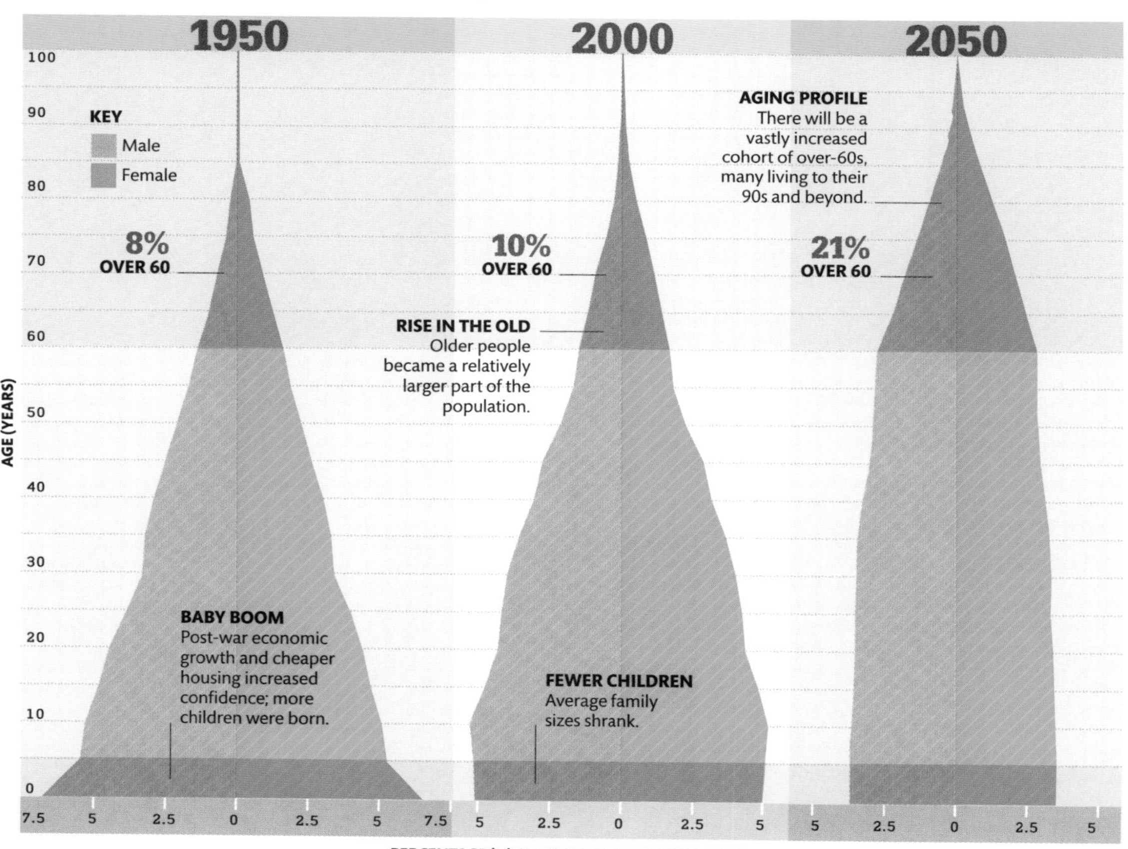

1950
The global population growth curve was steep. With an increase of nearly 19 per cent during the course of the 1950s, the high rate of growth persisted through the 1960s and 70s.

2000
The 50 years up to the new millennium saw the proportion of over-60s grow by 2 per cent. Declining fertility rates and changing causes of death heralded more rapid change ahead.

2050
Another demographic time bomb explodes. This one is not just about the overall increase in population, but includes a simultaneous doubling in the proportion of over-60s since 2000.

Slowing the rise

How best to manage population growth has been both one of the most discussed and most controversial questions of modern times. But what might actually work to reduce the rate of increase?

The steep population increase of the 20th century led to alarming predictions about its impact on the global environment, resources, and food supplies. While the humanitarian disaster some expected has thus far been avoided, there are still good reasons to reduce population growth.

A number of steps have been taken in pursuit of this goal, including forced sterilization (in India), greater access to contraception (many African countries), and a legal limit on family size (China, see box, right). Less controversial – and ultimately more successful – has been access to education, especially for girls and young women.

Women's education and birth rates

Generally speaking, literate women average two children per family, while those who cannot read or write often give birth to six or more. The situation can be self-perpetuating, as female children of illiterate women are less likely to receive education themselves.

Other benefits come with greater education. For example, families of women with at least some education tend to have better housing, clothing, income, water, and sanitation. Increased access to education thus emerges as a key area for investment, bringing social, economic, and – ultimately – environmental benefits.

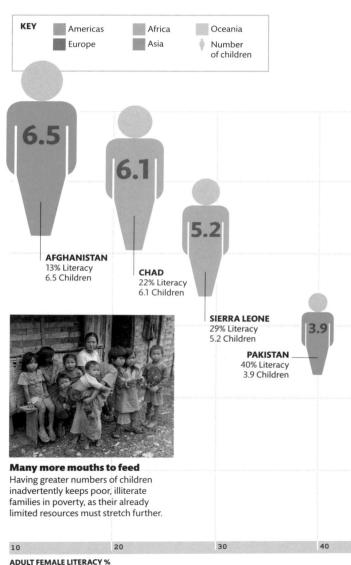

KEY
Americas
Europe
Africa
Asia
Oceania
Number of children

6.5
AFGHANISTAN
13% Literacy
6.5 Children

6.1
CHAD
22% Literacy
6.1 Children

5.2
SIERRA LEONE
29% Literacy
5.2 Children

3.9
PAKISTAN
40% Literacy
3.9 Children

10
20
30
40
ADULT FEMALE LITERACY %

Many more mouths to feed
Having greater numbers of children inadvertently keeps poor, illiterate families in poverty, as their already limited resources must stretch further.

The literacy link

On average, women who read and write have fewer children. Where there is a higher adult literacy rate but high numbers of children per woman, the result is often due to gender inequality – more men being literate than women.

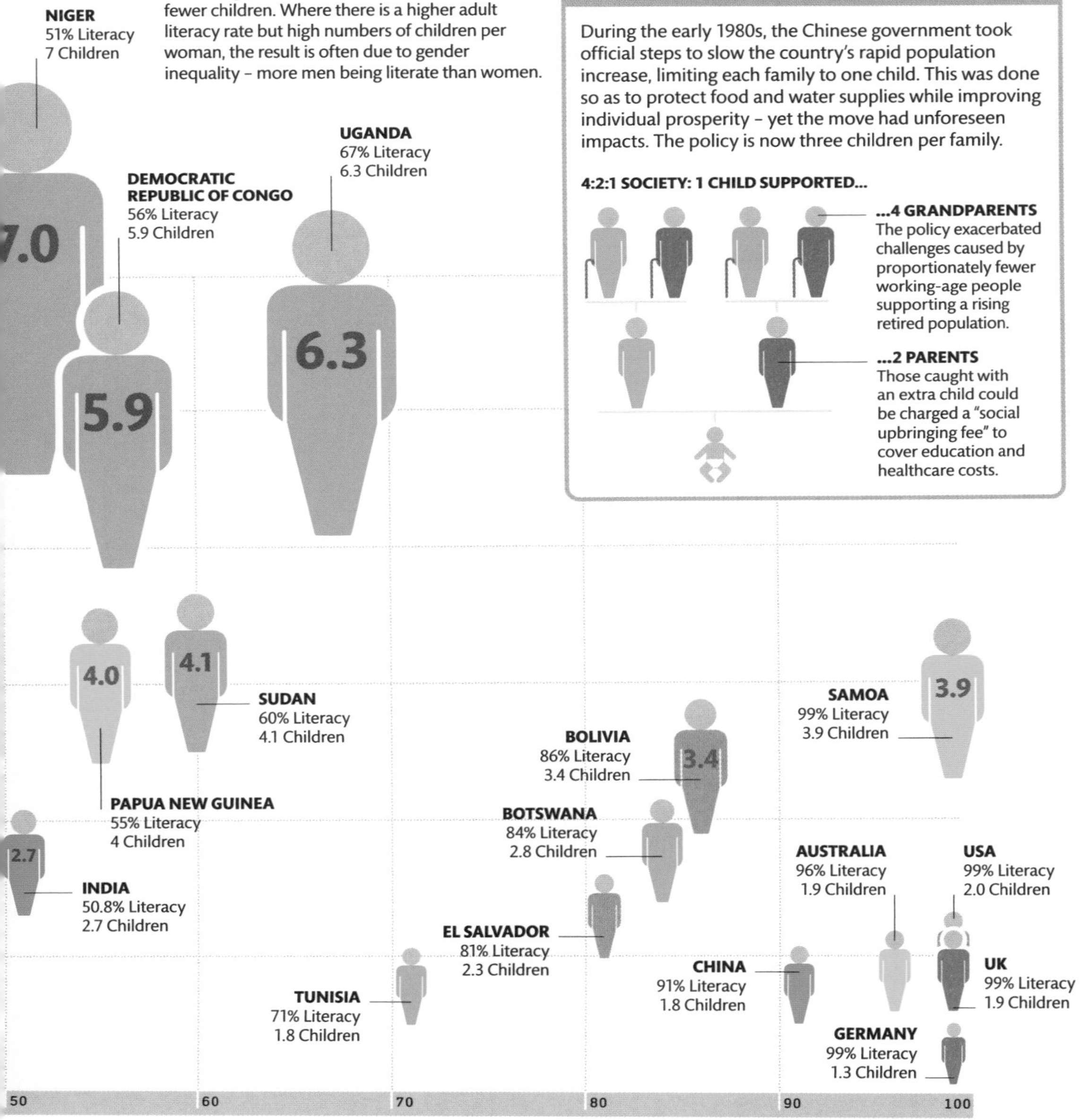

NIGER
51% Literacy
7 Children

7.0

DEMOCRATIC REPUBLIC OF CONGO
56% Literacy
5.9 Children

5.9

UGANDA
67% Literacy
6.3 Children

6.3

4.0

4.1

SUDAN
60% Literacy
4.1 Children

PAPUA NEW GUINEA
55% Literacy
4 Children

2.7

INDIA
50.8% Literacy
2.7 Children

BOLIVIA
86% Literacy
3.4 Children

3.4

SAMOA
99% Literacy
3.9 Children

3.9

BOTSWANA
84% Literacy
2.8 Children

AUSTRALIA
96% Literacy
1.9 Children

USA
99% Literacy
2.0 Children

EL SALVADOR
81% Literacy
2.3 Children

CHINA
91% Literacy
1.8 Children

UK
99% Literacy
1.9 Children

TUNISIA
71% Literacy
1.8 Children

GERMANY
99% Literacy
1.3 Children

50 60 70 80 90 100

CHINA'S ONE-CHILD POLICY

During the early 1980s, the Chinese government took official steps to slow the country's rapid population increase, limiting each family to one child. This was done so as to protect food and water supplies while improving individual prosperity – yet the move had unforeseen impacts. The policy is now three children per family.

4:2:1 SOCIETY: 1 CHILD SUPPORTED...

...4 GRANDPARENTS
The policy exacerbated challenges caused by proportionately fewer working-age people supporting a rising retired population.

...2 PARENTS
Those caught with an extra child could be charged a "social upbringing fee" to cover education and healthcare costs.

Economic expansion

Since the beginning of the Industrial Revolution in the late 18th century the world has witnessed a period of staggering economic growth. New methods of production and innovation developed over the last 200 years have allowed for efficient use of labour and resources, producing more output per person. Increased productivity has allowed for higher incomes, better quality of life, and greatly reduced poverty worldwide. As fast-growing countries in Asia, South America, and Africa progressively industrialize, the global economy is set to grow further still.

A more productive world

The total economic output of the world, its GDP, has been growing steadily, especially since the 1950s. Major drivers of economic growth are increasing populations, providing more workers to produce goods and services, cheap supplies of energy, and more advanced technologies, allowing labour to be more efficiently utilized. In 2015, global economic output was about 10 times its 1950 level. Even as growth slowed following the financial crisis of 2008-09, economic output reached new all-time highs in the ensuing decade.

"We have allowed the **interests of capital** to **outweigh the interests of human beings and our Earth.**"

ARCHBISHOP DESMOND TUTU, SOUTH AFRICAN SOCIAL RIGHTS CAMPAIGNER

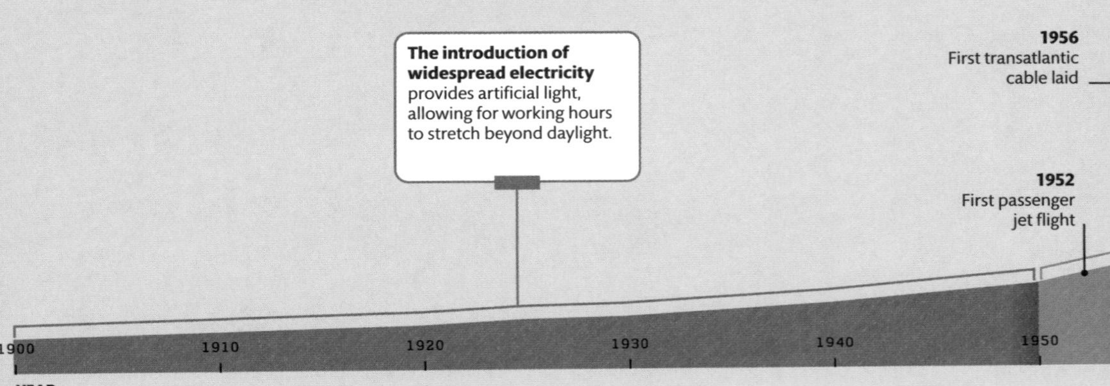
The introduction of widespread electricity provides artificial light, allowing for working hours to stretch beyond daylight.
1956
First transatlantic cable laid
1952
First passenger jet flight
1900
1910
1920
1930
1940
1950
YEAR

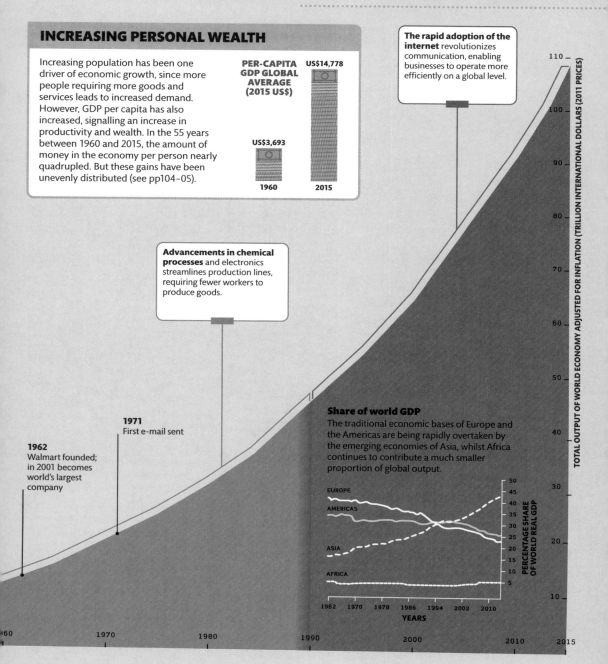

INCREASING PERSONAL WEALTH

Increasing population has been one driver of economic growth, since more people requiring more goods and services leads to increased demand. However, GDP per capita has also increased, signalling an increase in productivity and wealth. In the 55 years between 1960 and 2015, the amount of money in the economy per person nearly quadrupled. But these gains have been unevenly distributed (see pp104–05).

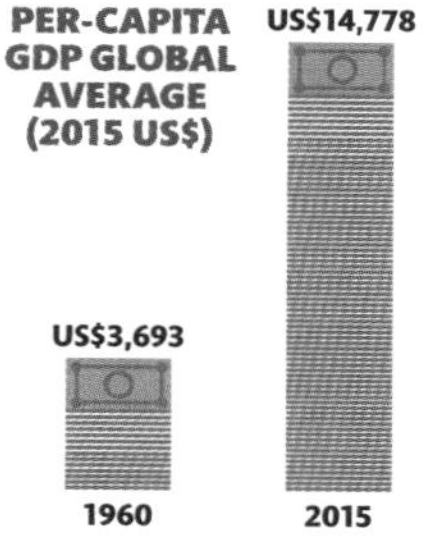

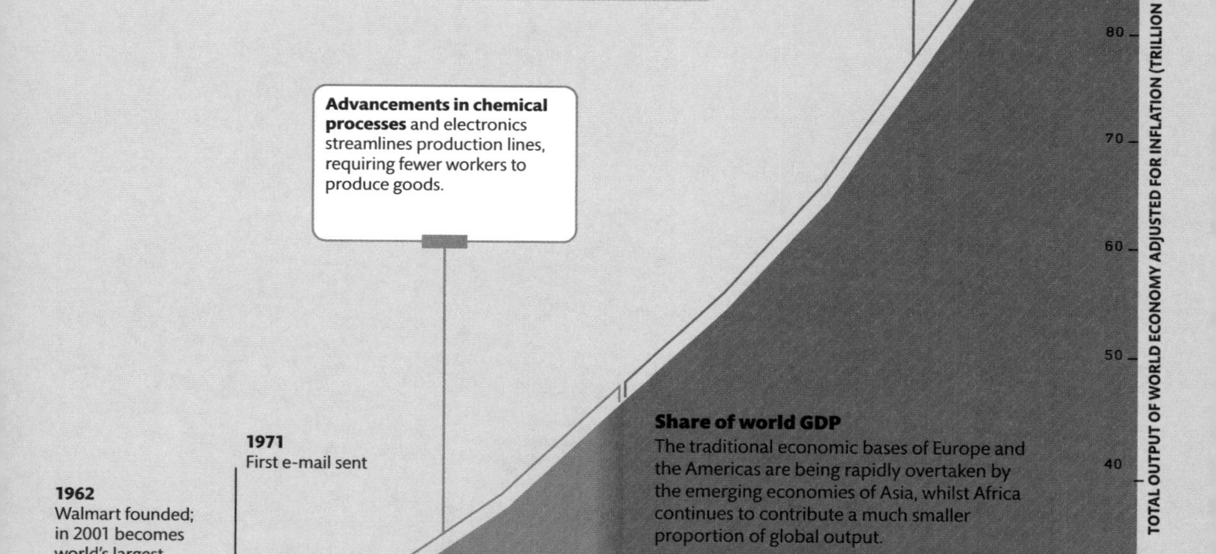

Share of world GDP

The traditional economic bases of Europe and the Americas are being rapidly overtaken by the emerging economies of Asia, whilst Africa continues to contribute a much smaller proportion of global output.

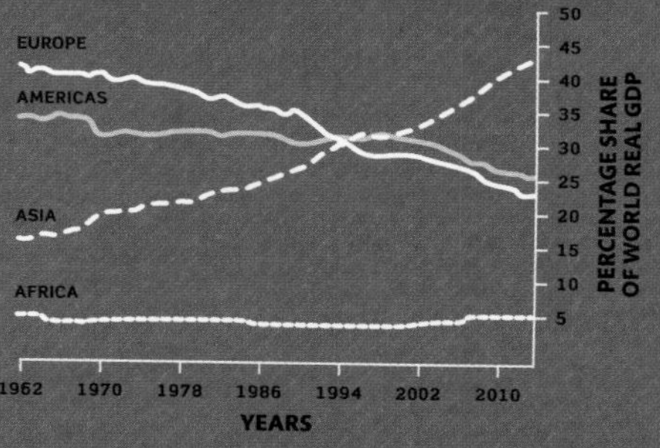

What is GDP?

Gross Domestic Product, or GDP, is a measure of the output of an economy, which is defined as the total value of all the finished goods and services produced within the borders in a specific time period, usually a year. It is used to compare the relative size of economies and to judge the health of an economy over time. There are several ways in which economists measure this output – here we look at the expenditure method. This assesses the output by adding up the total amount spent by the government, individuals, businesses, and organisations in the economy.

KEY

(C) Consumer spending
The total value of all goods and services bought by individuals and households

(I) Investment spending
Money spent by companies on equipment to enable them to provide goods and services in the future; new residential purchases

(G) Government spending
What the government spends on public services and public sector salaries

(X) Net exports
The value of goods and services the country produces and exports for sale in other countries, minus the value of imports

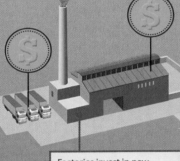

The government buys planes and weapons from production companies and pays the wages of soldiers and workers.

Factories invest in new equipment and machinery to produce goods for sale.

GDP

= C + I + G + X

There are several ways to calculate GDP. Here it is shown as the sum of expenditure in four components – consumer spending, investment spending, government spending, and net exports.

By trading with other countries, the economy can sell its domestically-produced goods and services abroad.

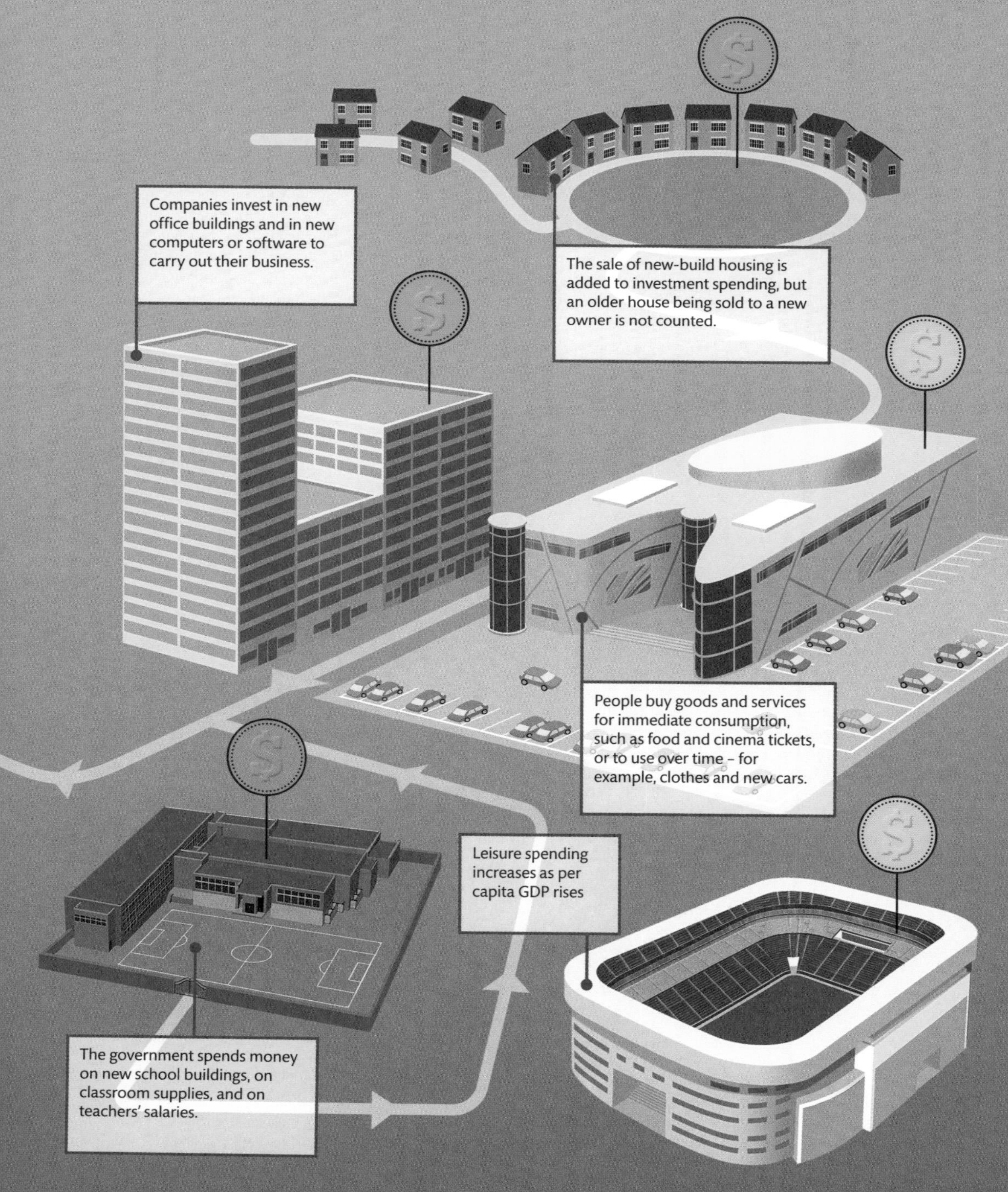
Companies invest in new office buildings and in new computers or software to carry out their business.
The sale of new-build housing is added to investment spending, but an older house being sold to a new owner is not counted.
People buy goods and services for immediate consumption, such as food and cinema tickets, or to use over time – for example, clothes and new cars.
Leisure spending increases as per capita GDP rises
The government spends money on new school buildings, on classroom supplies, and on teachers' salaries.

Richer people

Across the world, people are earning more money and can afford a better standard of living, but the gulf between the richest and poorest among us continues to grow wider.

A useful way of gauging how economic growth or decline is impacting on individual quality of life in different countries is to look at gross domestic product (GDP – see pp26–27) per capita. This is the measure of a country's annual economic output divided by its population. GDP per capita figures give an indication of individual average income and quality of life, allowing for comparisons over time to see if people are generally living better or worse lives. Globally, average GDP per capita rose from US$4,271 in 1990 to US$11,443 in 2019, signalling an overall rise in household earnings. This is in part due to the rise of emerging economies – such as Brazil, Russia, India, and China – and has led to significant reductions in poverty in some of the poorest countries in the world. However, by far the biggest factor in rising average GDP during this period is the continuing growth of the world's richest economies. Established economies, such as the USA and UK, may have slower growth rates but have much higher GDP per capita.

SEE ALSO...

- **Global power shift** pp30–31
- **Rise of consumerism** pp80–81
- **Unequal world** pp104–05

Global inequality

Despite the fact that global GDP per capita is increasing, and few countries have negative growth rates, the gap between rich and poor countries is growing. In the years 1990–2019, the most dramatic growth came in the emerging economies of China, Vietnam, and India. Vietnam's growth has given rise to a more than twentyfold increase in per capita GDP, while China has increased per capita GDP by an even greater proportion. These are successes, but they are outstripped in absolute terms by countries like the USA and Norway with their more stable, established economies.

KEY
Percentage growth of GDP per capita, 1990–2019
1990 GDP per capita
2019 GDP per capita

SLOW GROWTH
Despite slower growth than most, Japan has one of the highest living standards in the world

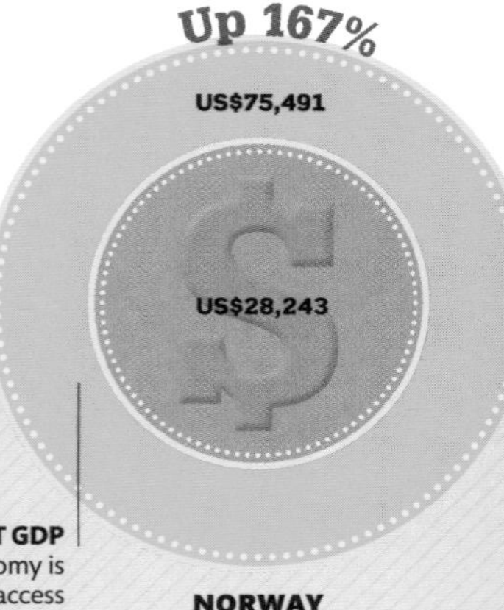

HIGHEST GDP
Norway's large economy is mostly due to their access to North Sea oil, which is in government ownership

MIDDLE CLASS WORLD

The global middle class, those with a daily spending power of $10–100, is expanding. Around 1.8 billion people were categorized as middle class in 2009 and this is expected to rise to 4.9 billion by 2030. The influence of middle-class consumers in the developing world is growing too. By 2030, it is estimated that around 35 per cent of global middle-class consumption will come from India and China.

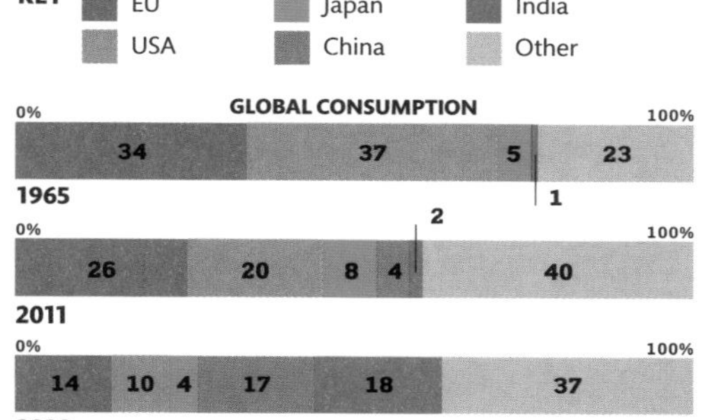

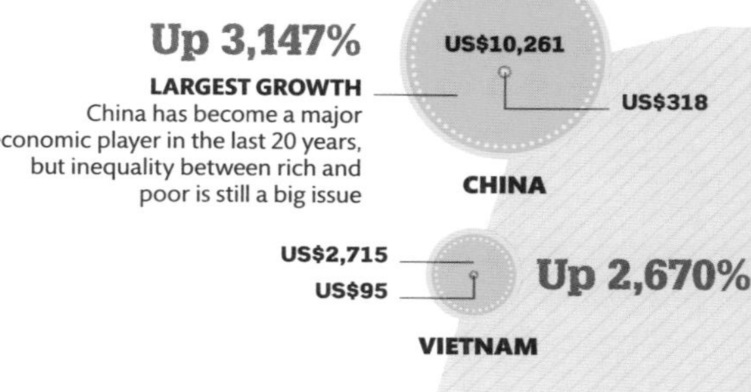

LARGEST GROWTH
China has become a major economic player in the last 20 years, but inequality between rich and poor is still a big issue

US$2,715
US$95
Up 2,670%
VIETNAM

INDIA

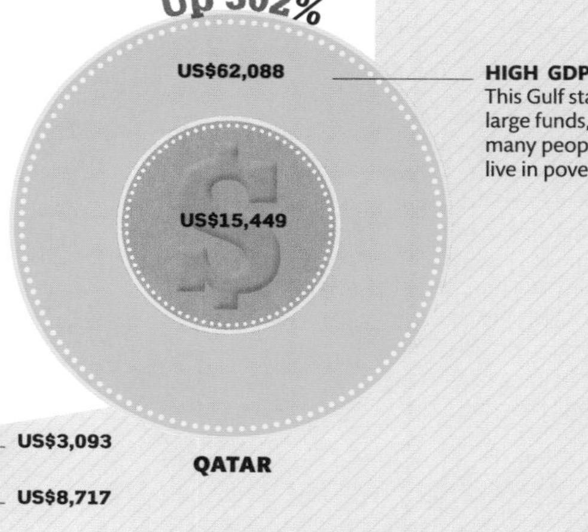

HIGH GDP
This Gulf state has large funds, but many people still live in poverty

QATAR

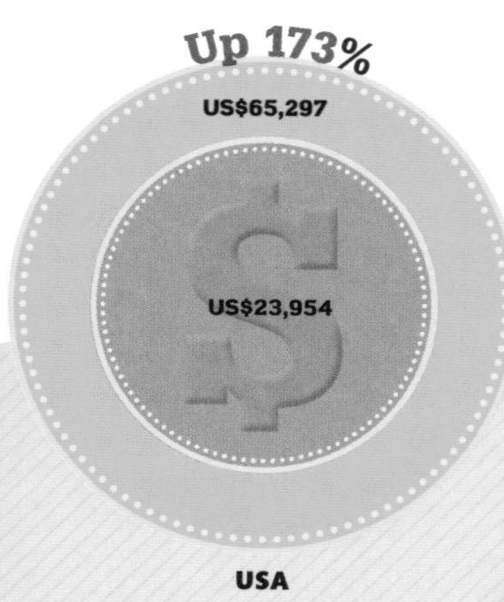

USA

Up 182%
US$3,093
US$8,717
BRAZIL

Conspicuous consumption
While China's GDP per capita has soared, the gap between rich and poor has grown. Only a small minority can afford luxuries such as this attention-grabbing Ferrari.

Global power shift

For the past 40 years, seven countries (the G7) have been accepted as the world's most important economies, but emerging economies are beginning to overtake them.

Since the late 19th century, the USA has been widely accepted as the world's largest economy and the leader in terms of output and innovation. Other traditional economic powerhouses joined with the USA to become the Group of 7, or G7, in the 1970s. The E7 group (or "Emerging 7"), identified in 2006, consists of the most important developing economies.

Growth of the E7

By 2050, the G7 economies are expected to be greatly overtaken by the seven emerging economies of the E7. In China, the reform of socialist economic policies and rapid expansion of manufacturing capabilities has led to huge economic expansion, which is predicted to continue. By 2050, India will also overtake the USA and become the world's second largest economic power. The G7 economies will continue to grow, but at a much slower rate than their emerging counterparts.

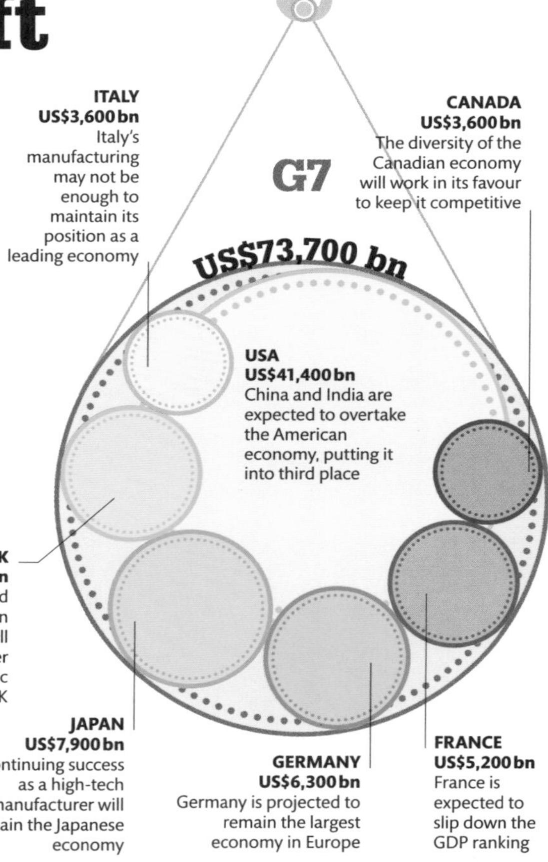

ITALY
US$3,600bn
Italy's manufacturing may not be enough to maintain its position as a leading economy

CANADA
US$3,600bn
The diversity of the Canadian economy will work in its favour to keep it competitive

USA
US$41,400bn
China and India are expected to overtake the American economy, putting it into third place

UK
US$5,700bn
Expected population increases will be one driver of economic growth in the UK

JAPAN
US$7,900bn
Continuing success as a high-tech manufacturer will sustain the Japanese economy

GERMANY
US$6,300bn
Germany is projected to remain the largest economy in Europe

FRANCE
US$5,200bn
France is expected to slip down the GDP ranking

THE WORLD'S RICHEST 50 CITIES

The economic rise of the East is well illustrated by forecasts of where we will soon find the world's richest cities. In 2007, eight of the richest 50 cities ranked by annual GDP were in Asia. By 2025, this is predicted to rise to 20. More than half of Europe's top 50 cities are expected to drop off the list entirely, as will three in North America, creating a new landscape of urban economic power.

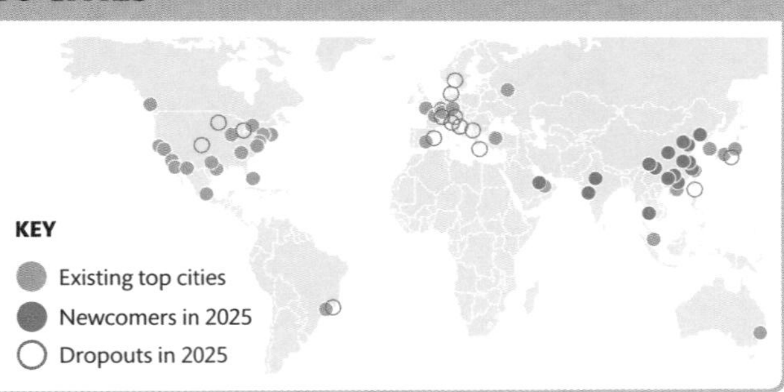

The EU and USA's share of world GDP is expected to fall from 33% in 2014 to around 25% by 2050

Tipping the balance
An economic power shift away from the established advanced economies in North America, Western Europe, and Japan is predicted to continue. By 2050, the combined GDP of the E7 nations is expected to be double that of the G7.

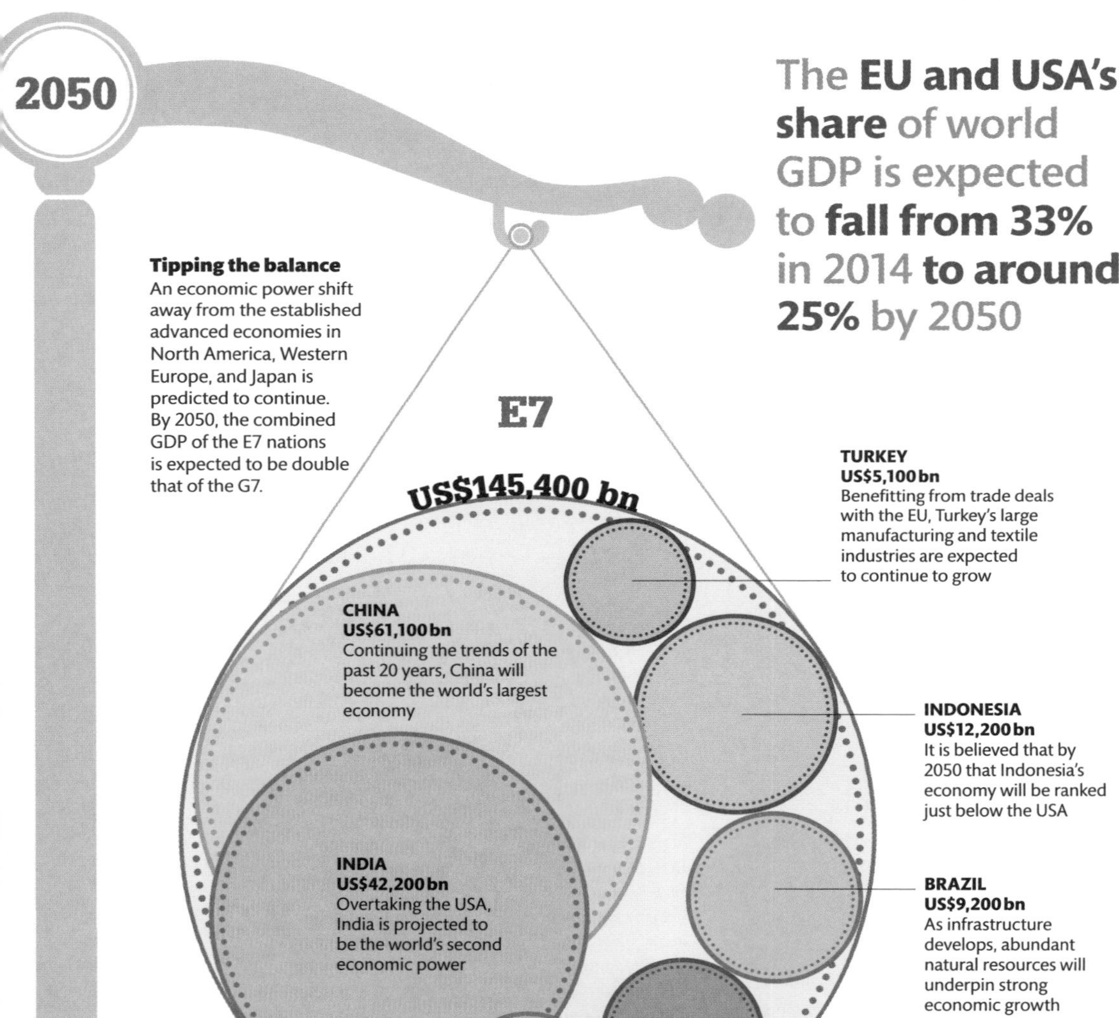

CHINA
US$61,100 bn
Continuing the trends of the past 20 years, China will become the world's largest economy

INDIA
US$42,200 bn
Overtaking the USA, India is projected to be the world's second economic power

TURKEY
US$5,100 bn
Benefitting from trade deals with the EU, Turkey's large manufacturing and textile industries are expected to continue to grow

INDONESIA
US$12,200 bn
It is believed that by 2050 that Indonesia's economy will be ranked just below the USA

BRAZIL
US$9,200 bn
As infrastructure develops, abundant natural resources will underpin strong economic growth

MEXICO
US$8,000 bn
Mexico is projected to continue to supply it's North American neighbours with 90 per cent of its exports, providing a source of sustained income

RUSSIA
US$7,600 bn
Russia's diverse natural resources will continue to be a major export and driver of economic success

Trading benefits

Trade has been a powerful driver of economic growth around the world for centuries. Countries described as major traders have larger economies than the smaller trading nations.

Trade enables countries to make the most of their natural and human resources. Modern transport is now so fast and efficient that even perishable foods and flowers can be harvested in southern Africa and sold in European supermarkets within days. The use of instant internet communications means that many services are no longer restricted by location. These technological advances have resulted in a boom in the value of international trade.

World trade

The majority of international trade (measured here as total exports) takes place between the richest countries. They benefit from efficient infrastructure and supportive treaties, and are able to produce goods of high value. The ease of trade and transport today means that almost any product or service is available across the world.

TRADE VERSUS AID

Some experts believe international aid should be reduced in favour of investment in trade with poorer countries to support their development.

Trade

- Establishes a partnership rather than a one-way dependent relationship
- Fosters development of industry and infrastructure in poorer countries
- Can leave countries heavily dependent on powerful foreign countries

Aid

- Provides relief and support in a crisis
- Can be used to encourage policies for sustainable development
- Foreign aid can leave economies unequipped and dependent on assistance

Least developed countries

The 48 least developed countries, as set by the UN, are impeded in trade by a lack of infrastructure and supportive government. Goods and services of low value are often traded here.

IMPORTS
Lack of manufacturing capacity in many poorer countries means they cannot participate in key global markets. These nations must import manufactured goods, such as vehicles and medicines.

EXPORTS
The leading exports of many less developed countries are often natural resources, used abroad to produce manufactured goods. Tourism brings in income as a service export.

LABOUR
The countries concentrated on extracting raw materials may suffer from so-called "Dutch disease", whereby exporting raw materials is at the expense of jobs in more stable or lucrative manufacturing industries.

US $236 billion
Least developed countries

S $23.6 trillion World trade

US $23,300 billion Rest of the world

90% of world trade is carried by the shipping industry

Developed countries

Trade agreements and open borders often make it cheaper for groups of richer countries to trade together. Good infrastructure and communication links ensure trade is easy to conduct.

IMPORTS
Food, raw materials, and machinery are all imported regularly to produce manufactured goods. Rich countries can afford to import basic goods and services, allowing them to specialise in high-value industries.

EXPORTS
The highest-value exports of many developed countries are consumer electronics and vehicles. Services are exported in the form of financial services and travel, as well as the tourism industry.

LABOUR
Many large economies, such as China and the USA, produce large amounts of consumer goods for export. This supports millions of skilled jobs in these countries.

American agreement

The USA is the world's largest international trader, with trade valued at more than US$3,900 billion in 2014. With the North American Free Trade Agreement (NAFTA), the USA's biggest partner is Canada. A third of US exports go to Canada and Mexico.

CANADA
Trade with Canada is vital for both economies, and the two countries share the highest-value trade relationship in the world.

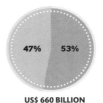

47% 53%
US$ 660 BILLION

CHINA
The USA's largest source of imports come from China. Exports are also growing rapidly, making China the third biggest market for US goods and services abroad.

20% 80%
US$ 590 BILLION

MEXICO
The third member of the NAFTA, Mexico has cheaper labour and production costs. This means many consumer goods are exported to the USA.

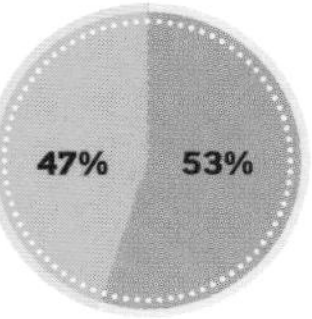

US$ 534 BILLION

JAPAN
Imports from Japan are almost exclusively manufactured goods, with cars and electronics the most popular items.

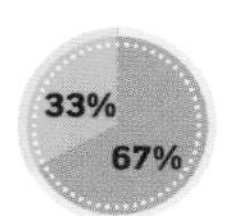

US$ 201 BILLION

GERMANY
The USA's largest European trading partner is Germany, known for exporting high-quality consumer goods.

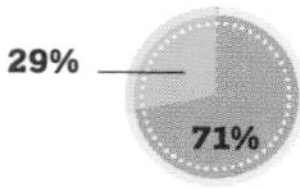

US$ 173 BILLION

City planet

The first organized urban centres were founded over 10,000 years ago. They arose in parallel with agricultural advances that enabled farmers to produce the food surpluses required to feed the new urban populations. Urbanization gathered pace with the Industrial Revolution and the intensive agriculture that enabled farmers to produce more food. Urban migration continues to increase, but so do concerns about its sustainability. By 2050 new urban capacity equivalent to 175 times the size of London will be needed to accommodate town and city dwellers.

Rural-to-urban shift

In 1800 about two per cent of the world's population lived in urban areas. Over time, millions of people who once farmed have moved to urban areas in search of better lives, or have been forced to move because of falling incomes. In 2007, for the first time, more than half of us lived in towns and cities. Continuing population growth and urbanization are projected to add 2.5 billion people to the world's urban population by 2050. This works out as about 180,000 people every day, mostly in fast-growing developing countries. In 2019 the global urban population was nearly 4.3 billion.

> "**In many cities** the strain on both infrastructure (housing, water, sewerage, transport, electricity supply) and **the quality of life ... is becoming unbearable.**"
>
> **GEORGE MONBIOT, UK WRITER AND CAMPAIGNER**

1892
The Masonic Temple in Chicago, USA is the world's tallest skyscraper. Skyscrapers changed the way cities were built. The population of Chicago more than tripled from 1850–1900.

1920s
Social mixing during World War I encouraged many young people to migrate to urban areas in the years that followed the war.

1950s
Just 30 per cent of the world's total population lived in urban areas during the 1950s.

1890 1900 1910 1920 1930 1940 19

YEAR

UNEVEN URBANIZATION

In some countries, urban growth is nearly double the rate of overall population increase, particularly in the urban areas of less developed regions. Europe, North America, and Oceania have all experienced stable rates of urbanization in the last 15 years, while South America has witnessed continuously decreasing rates. Africa and Asia, meanwhile, are responsible for bringing the developing world's average up in recent years, with Africa expected to be the fastest-urbanizing region from 2020–2050.

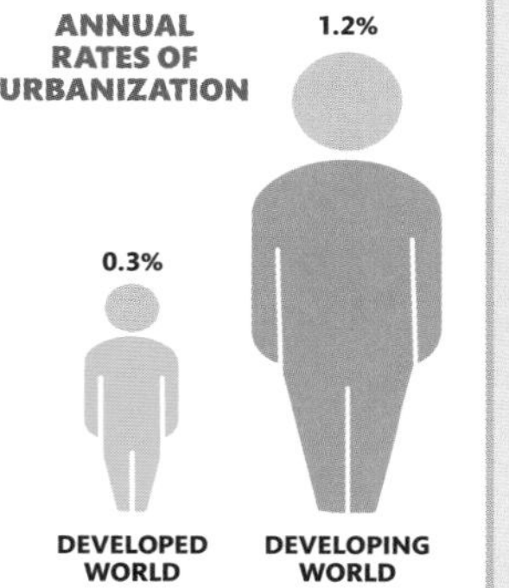

2007
In 2007 the historic point was reached when more than half the world's population lived in towns and cities.

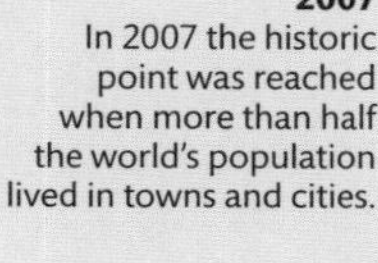

Industrialization, more intensive farming, and new infrastructure facilitate an unprecedented period of urbanization.

1980s
The 1980s sees a rapid growth in the urban population, including in China.

Developing trends
Africa and Asia remain mostly rural, but they are urbanizing faster than other continents. Their urban proportions of their total populations are expected to rise to 56 and 64 per cent, respectively, by 2050.

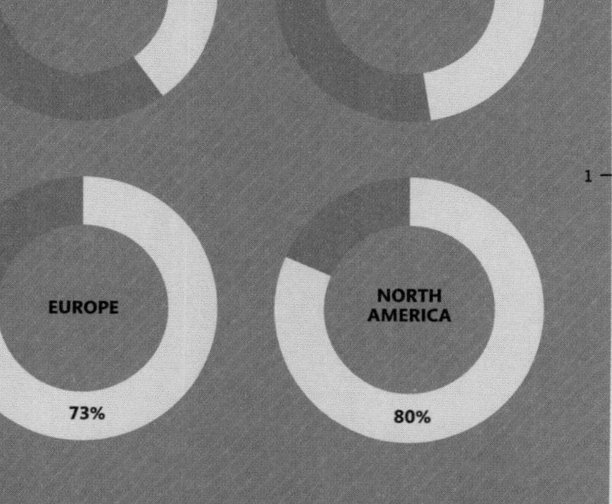

KEY
Proportion of total population (per cent 2014)
Urban
Rural

The rise of megacities

The past 25 years has seen rapid growth in the number of megacities – cities with a population of over 10 million. In 1950 there was only one in the world – New York City. By 1990 there were 10. This number has more than tripled to 31 today.

In recent decades, the centres of world urbanization have shifted from the developed countries of Japan, North America, and Europe to the developing nations of Asia, Africa, and South America.

This shift is reflected in the United Nations projection that by 2030 there will be another 10 megacities, all of which are in developing countries. These new megacities are anticipated to be Lahore, Hyderabad, Bogota, Johannesburg, Bangkok, Dar es Salaam, Ahmedabad, Luanda, Ho Chi Minh City, and Chungdu.

Africa is experiencing rapid urbanization. For example, Kinshasa, in the Democratic Republic of the Congo, will see its population rise from 200,000 in 1950 to a projected 20 million in 2030, up from around 12 million in 2016. Some megacities will be poorly prepared for such rapid growth, placing great strain on natural resources, food, and transport.

SEE ALSO...

- **Global power shift** pp30–31
- **Rise of consumerism** pp80–81
- **Unequal world** pp104–05

Changes in the 10 largest cities

Asia has already seen spectacular growth, with 11 of the 31 cities that now exceed 10 million people located in China and India alone. However, not all of Asia is growing at such a rapid rate. Rising life expectancy and a relatively low birth rate will have a profound impact on Japan. Tokyo is currently the largest megacity and will continue to be in 2030, but Delhi is catching up.

In 1990 there were **10 cities with more than 10 million inhabitants.** Today, the total has **more than tripled**

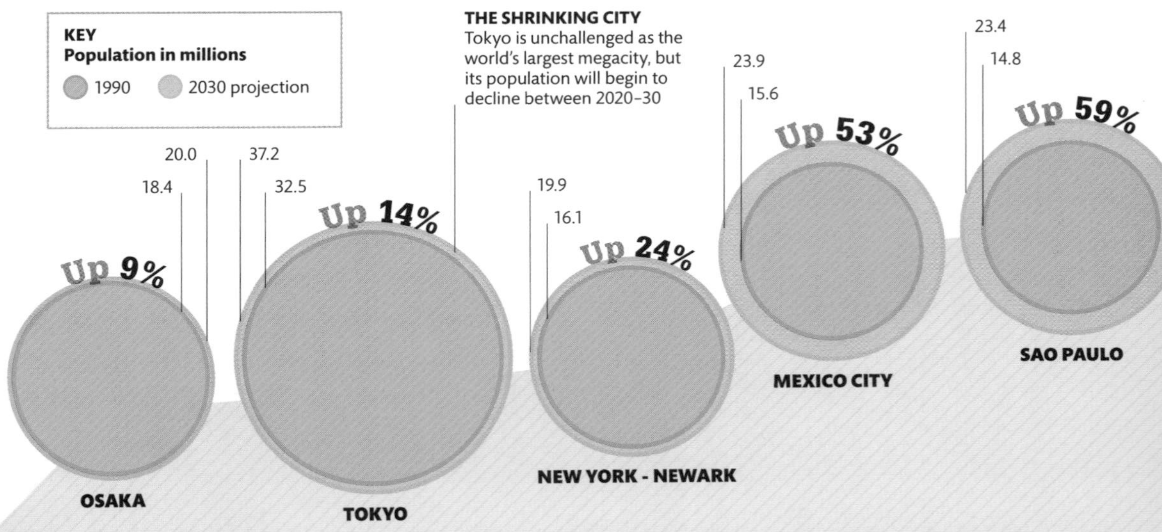

Up 308%
27.7
6.8
BEIJING

Up 293%
30.7
7.8
SHANGHAI

Up 271%
36.1
9.7
DELHI

NEW CHALLENGE Delhi's population is expected to nearly quadruple, challenging Tokyo's dominance as the largest megacity

STRATOSPHERIC RISE If the projections are realized, Beijing will be the fastest growing megacity, living up to its nickname of the Celestial City.

PUSHED INTO SECOND Mumbai's population is projected to more than double, but it will no longer be India's largest city

Up 148%
24.5
9.9
CAIRO

Up 124%
27.8
12.4
MUMBAI

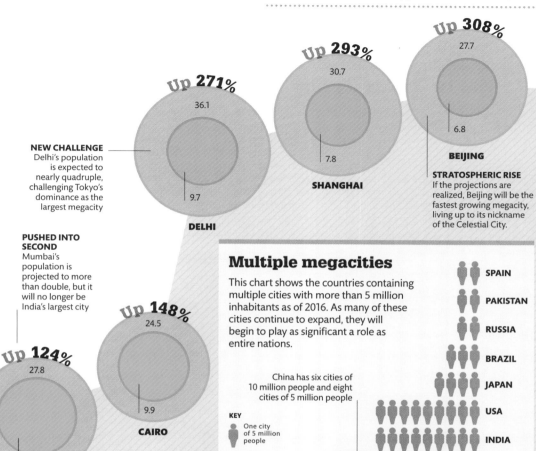

Multiple megacities

This chart shows the countries containing multiple cities with more than 5 million inhabitants as of 2016. As many of these cities continue to expand, they will begin to play as significant a role as entire nations.

China has six cities of 10 million people and eight cities of 5 million people

KEY
One city of 5 million people
One city of 10 million people

SPAIN
PAKISTAN
RUSSIA
BRAZIL
JAPAN
USA
INDIA
CHINA

DISTRIBUTION OF MEGACITIES

The current distribution of the 31 megacities is concentrated strongly in Asia. There are now 18 megacities in Asia, four in South America and three each in Africa, Europe, and North America. Given that only 48 per cent of people in Asia live in cities, and this is expected to rise to 64 per cent by 2050, the number of megacities in this part of the world will continue to rise. Pressure on finite resources will be unprecedented.

Urban pressures

People living in cities tend to consume more energy, water, food, and resources than those in rural areas. Urban populations are responsible for about three-quarters of total consumption and half of all waste.

Cities are economic engines. Fuelled by natural resources, they generate most of the activity that leads to growth and wealth creation. This in turn, leads to more people migrating from rural areas to the cities, which brings with it some disadvantages. The increase in the number of city dwellers requires more food, water, and energy. The use of private and public transport also increases and more pollution is produced. Often, former rural dwellers adopt higher-consumption lifestyles in the cities, further increasing the demand for natural resources. All these factors can lead to the destruction of natural habitats and damage the environment through increased consumption.

URBAN DENSITY

Cities vary hugely in population density. An interesting way to compare urban density is to consider how large a city would need to be to accommodate all 7.3 billion people in the world, concentrated at the same rate. A city with the population density of New York would fit neatly into the state of Texas – an area of 648,540 km^2 (250,400 sq. miles), whereas a city with the low population density of Houston would occupy most of the landmass of the USA at 4,581,910 km^2 (1,769,085 sq. miles). Paris has a population density four times that of London.

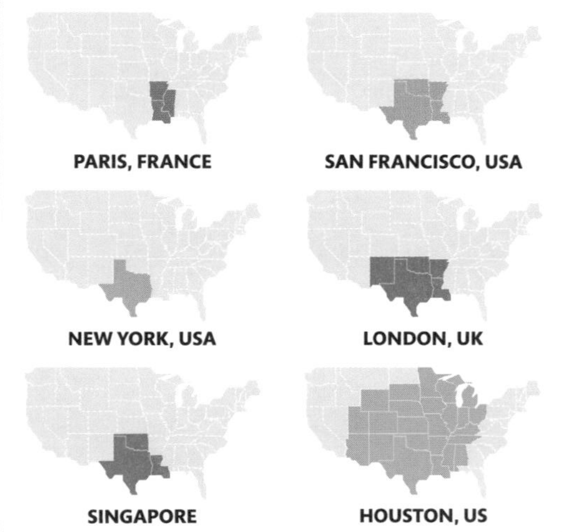

Ecological footprints

An ecological footprint measures the impact of human activities on the natural environment. It is essentially an area measurement, represented in global hectares, which places a value on how much biologically productive land and water is needed to both produce the resources we consume and to dispose of the waste. Every person, activity, company, and country has an ecological footprint. London's ecological footprint was analysed as part of a report entitled "City Limits". Published in 2002, it outlined the changes needed to turn London into a sustainable city.

2% of the world's land surface is occupied by cities, which **consume 75% of the world's natural resources**

LONDON'S ECOLOGICAL FOOTPRINT (2000)
At 293 times the size of London's geographical footprint, its ecological footprint is 49 million global hectares (gha) – equivalent to the area of Spain. London's population in 2000 was 7.4 million people.

44%
MATERIALS AND WASTE
The biggest part of London's ecological footprint was in the consumption of 49 million tonnes (54 million tons) of materials. The construction sector consumed the most materials at 27. 8 million tonnes (30.6 million tons) and also produced the most waste at 14.8 million tonnes (16.3 million tons).

LONDON'S GEOGRAPHICAL FOOTPRINT
The physical area covered by London measures 170,680 hectares or 1,706 km^2 (659 sq. miles).

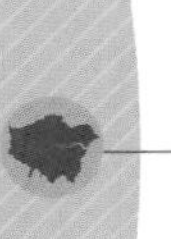

41%
FOOD
The consumption of 6.9 million tonnes (7.6 million tons) of food comprised the second largest part of London's footprint. Of the total food consumed, 81 per cent was imported from outside the UK. By far the largest component in the food ecological footprint was meat, followed by pet food and milk.

10%
ENERGY 10%
Londoners consumed energy equivalent to that present in 13.3 million tonnes (14.6 million tons) of oil, which in turn led to the release of about 41 million tonnes (45 million tons) of CO_2.

0.3%
WATER
London used 866,000 megalitres (190 billion gallons) in 2002, half of which was piped to houses. Water lost through leakage (about a quarter) was more than that used by businesses.

5%
TRANSPORT
Londoners travelled over 64 billion passenger km (40 billion miles), 44 billion (27 billion miles) of which were by car and light truck. Transport caused 8.9 million tonnes (9.8 million tons) of CO_2 emissions.

0.7%
DEGRADED LAND
This is land that has had its bioproductivity degraded through contamination or erosion, including roads, runways, and tracks.

Fuel for growth

Since our ancestors' first use of fire, the human race has continually sought access to ever-more-diverse energy sources. For centuries, economic development depended on the energy provided by animals, wood, wind, and water. Today, however, we rely on access to vast quantities of fossil energy from oil, coal, and gas to fuel electricity generation and power manufacturing, industrial farming, and long-distance transportation, as well as to drive the higher-consumption lifestyles that have developed as a result of each of these activities.

The energy revolution

The 20th century saw a massive surge in demand for energy that continues today, with the emergence of major economies such as China, India, Brazil, and South Africa. Meanwhile, other energy types have more recently played important roles, including nuclear power, hydropower, and modern technologies that harness energy from the wind and the Sun. Meeting any rising future demand presents a range of challenges, including those that relate to affordability and reducing climate-changing emissions and air pollution.

"We can **no longer continue feeding our addiction to fossil fuels** as if there were no tomorrow. **For there will be no tomorrow."**

ARCHBISHOP DESMOND TUTU, SOUTH AFRICAN HUMAN RIGHTS CAMPAIGNER

With the first industrial revolution, which began in 1750, water and steam create the first mechanical textile factories. Farming output almost doubles.

Energy from fossil fuels enables cheap mass production, the spread of manufactured fertilizers, and the rise of entire new industries.

1882
Pearl Street power station opens in New York. The first to use coal, it heralds the rise of mass electricity use

1914
World War I begins and signals the rising dominance of oil for transport

1840 1860 1880 1900 1920

YEAR

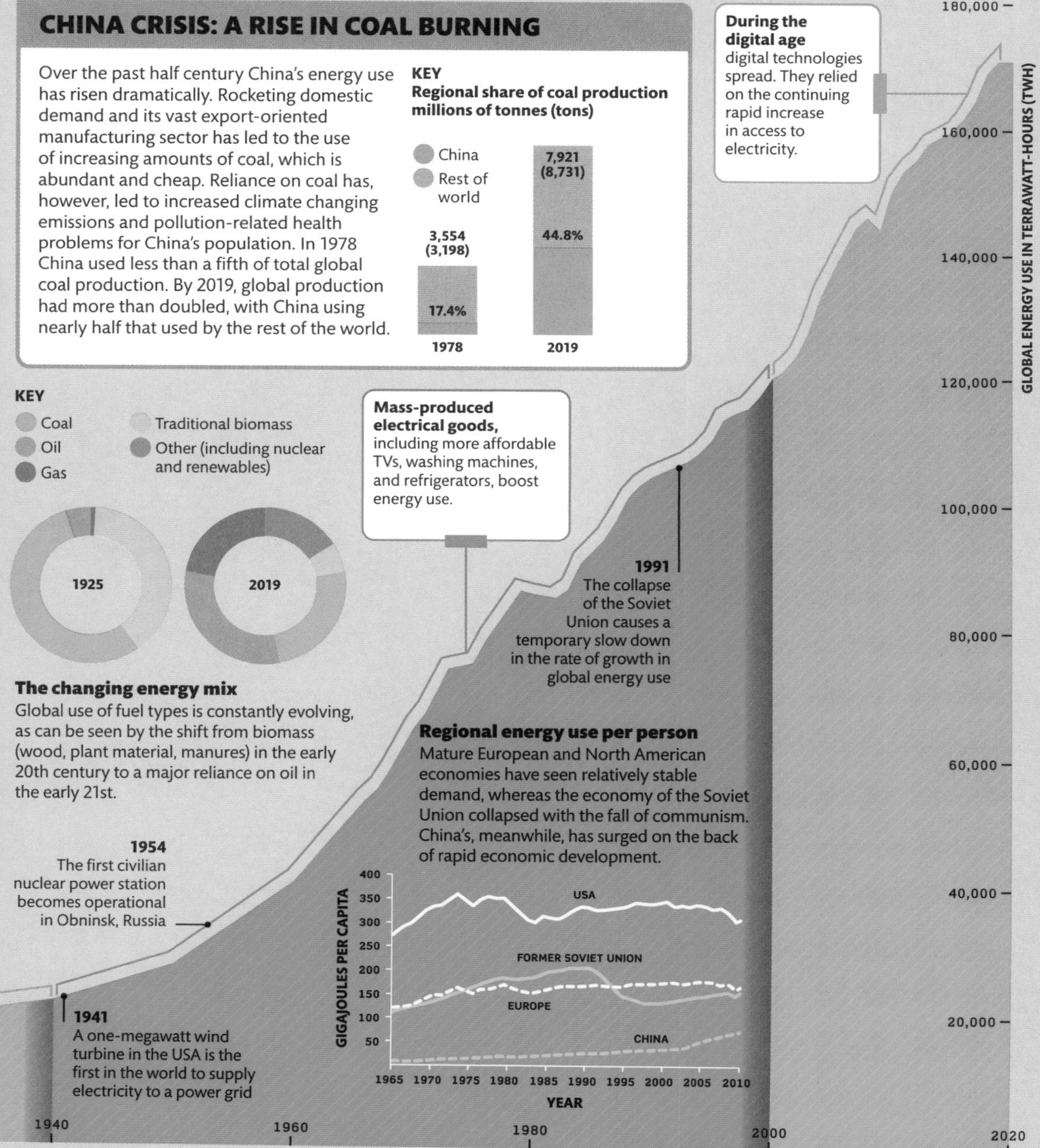
CHINA CRISIS: A RISE IN COAL BURNING
Over the past half century China's energy use has risen dramatically. Rocketing domestic demand and its vast export-oriented manufacturing sector has led to the use of increasing amounts of coal, which is abundant and cheap. Reliance on coal has, however, led to increased climate changing emissions and pollution-related health problems for China's population. In 1978 China used less than a fifth of total global coal production. By 2019, global production had more than doubled, with China using nearly half that used by the rest of the world.
KEY
Regional share of coal production millions of tonnes (tons)
China
Rest of world
3,554 (3,198)
17.4%
1978
7,921 (8,731)
44.8%
2019
During the digital age
digital technologies spread. They relied on the continuing rapid increase in access to electricity.
KEY
Coal
Oil
Gas
Traditional biomass
Other (including nuclear and renewables)
1925
2019
The changing energy mix
Global use of fuel types is constantly evolving, as can be seen by the shift from biomass (wood, plant material, manures) in the early 20th century to a major reliance on oil in the early 21st.
Mass-produced electrical goods,
including more affordable TVs, washing machines, and refrigerators, boost energy use.
1991
The collapse of the Soviet Union causes a temporary slow down in the rate of growth in global energy use
Regional energy use per person
Mature European and North American economies have seen relatively stable demand, whereas the economy of the Soviet Union collapsed with the fall of communism. China's, meanwhile, has surged on the back of rapid economic development.
1954
The first civilian nuclear power station becomes operational in Obninsk, Russia
1941
A one-megawatt wind turbine in the USA is the first in the world to supply electricity to a power grid
GIGAJOULES PER CAPITA
400
350
300
250
200
150
100
50
USA
FORMER SOVIET UNION
EUROPE
CHINA
1965 1970 1975 1980 1985 1990 1995 2000 2005 2010
YEAR
180,000
160,000
140,000
120,000
100,000
80,000
60,000
40,000
20,000
GLOBAL ENERGY USE IN TERRAWATT-HOURS (TWH)
1940
1960
1980
2000
2020

Surge in demand

Economic growth has depended on access to vast quantities of cheap energy to generate electricity, produce heat, and for transport. Further development and urbanization mean demand will continue to rise.

Based on current figures, most of this projected increase is expected to occur in the fast-growing economies of the global East and South, such as Asia and Africa. It is believed that fossil energy will continue to make the greatest contribution towards meeting the world's rising demand.

In the past, the human world was powered mainly by renewable energy in the form of wood, water, wind, and animal power. Since industrialization, we have increasingly relied on fossil fuels, and, to a limited extent, nuclear power. The increased use of natural gas to generate power (at the expense of coal) is helping to curb emissions to levels that are below what they would have been otherwise. However, it is clear that if we are to prevent global warming by limiting the average planetary temperature increase to below 1.5°C (2.7°F) compared with the pre-industrial period, then we will need to see a much lower reliance on fossil fuel sources and a much faster growth in renewable energy technologies.

SEE ALSO...

- **Carbon crossroads** pp130–31
- **Renewable revolution** pp46–47
- **Toxic air** pp136–37

Energy usage: present

The world's demand for energy continues to rise. By 2030, the amount of energy we need is expected to be about double the demand in 1990 and a third greater than that used in 2015. Today some countries are maintaining economic growth without causing rising emissions, but global demand for all energy types is increasing.

KEY

RENEWABLES
This category includes wind, solar, wave, tidal, and geothermal technologies. Some are still at small scale but growing fast.

BIOENERGY
Includes wood, sugar cane, and agricultural byproducts used as fuel sources to power transport and generate electricity and heat.

HYDRO ENERGY
Hydroelectric dams already generate substantial quantities of relatively low-carbon power. Expansion, however, is limited.

NUCLEAR ENERGY
Low-carbon at point of generation, this power source is expensive, with many technological and waste-management challenges.

NATURAL GAS
Although cleaner than coal, demand for gas is not compatible with strategies for limiting climate-changing emissions.

OIL
Used mainly to fuel road, sea, and air transport. Demand can be reduced through more efficient technology and electric vehicles.

COAL
By far the "dirtiest" power source, coal has played a major role in the development of many fast-growing countries, such as China and India.

TOTAL IN MILLION TONNES OF OIL EQUIVALENT (MTOE)
8,789

36 MTOE
905 MTOE
184 MTOE
526 MTOE
1,672 MTOE
3,235 MTOE
2,231 MTOE

1990

708 MTOE

1,827 MTOE

482 MTOE

1,044 MTOE

3,547 MTOE

40% Amount of all energy currently **used to produce electricity**

4,313 MTOE

3,448 MTOE

2030

The future of energy

Projections for future energy use show that while renewables will continue to grow and provide a larger proportion of energy, there will still be a reliance on high-polluting fuels such as oil and coal. Renewables face various challenges, however. For example, hydroelectric power is at risk from droughts caused by climate change, while energy-storage technologies still need to be refined to cope with the intermittency of some renewables.

What can we do?

- **Governments and international agencies** can use policies to create a faster transition to cleaner energy sources, while encouraging more efficient energy use among the industries that are the biggest consumers.
- **Governments** can shift public subsidies away from fossil fuel production towards cleaner, renewable energy alternatives.

What can I do?

- **Buy electricity** from companies generating power via renewable sources.
- **Reduce your energy use** Turn down heating, use air conditioners less, unplug unused appliances, switch off unneeded lights. Walk and cycle whenever possible.

Carbon footprint

Many of the things we do generate a carbon footprint. This footprint describes the quantity of carbon dioxide (CO_2) emissions arising from particular products, activities, or services.

Carbon footprints vary hugely. For example, that of an average American citizen is more than a hundred times that of a poor person living in sub-Saharan Africa. Some activities, such as a flight on an airliner, have a major short-term footprint, whereas other carbon-heavy decisions, such as buying a new car, will be spread over years and dependon how much the car is driven. Carbon footprints can be difficult to calculate very precisely, but still give a helpful indication of where the biggest impacts arise. This enables choices to be made by people, companies, and governments in limiting emissions.

Personal footprint

The average carbon footprint of a UK citizen is a total of around 10 tonnes (11 tons) of emissions per person per year. This chart shows the average carbon dioxide generated per person per year in the UK in 2005 from activities and products, not including non-energy-related CO_2 emissions and other greenhouse gas emissions.

KEY

Carbon dioxide (tonnes/tons)

CO_2 quantity of carbon dioxide emitted as a result of the activity specified.
CO_2e carbon dioxide equivalent. Carbon dioxide plus other greenhouse gases emitted, converted to the common unit of carbon dioxide (not included in total).

Watching TV for an hour on a 24-inch plasma screen **220 g (8 oz) CO_2e**
A trip to the gym **9.5 kg (21 lbs) CO_2**
Buying a CD album online **400 g (14 oz) CO_2**

0.52
(0.57)
RECREATION AND LEISURE
(All leisure activities from watching television to holidays, but excluding flying)

0.37
(0.41)
FOOD AND CATERING
(Agriculture, food, transportation, cooking, restaurants)

Cappuccino **235 g (8 oz) CO_2e**
1 kg (2 ¼ lbs) lamb **39.2 kg (86 lbs) CO_2e**
1 kg (2 ¼ lbs) chicken **6.9 kg (15 lbs) CO_2e**

A T-shirt from manufacture to disposal **10 kg (22 lbs) CO_2**

CLOTHING
(Production, road transportation, retail, and washing/drying of clothes and shoes)
0.27
(0.3)

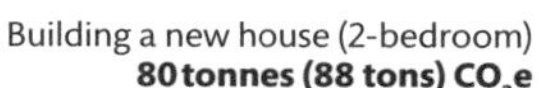

Building a new house (2-bedroom) **80 tonnes (88 tons) CO_2e**

HOUSEHOLD
(Including lighting, do-it-yourself, decoration, gardening)
0.37
(0.41)

Standard 100W light bulb **63 kg (140 lbs) CO_2 per year**
Lawn mower **73 kg (160 lbs) CO_2 per acre per year**

SPACE HEATING
(All forms of heating at home and at work)
0.4
(0.44)

COMMUTING
(Travelling to and from work by car or public transport)
0.22
(0.24)

Bus trip **66 g CO_2e per passenger km/3 ¾ oz CO_2e per passenger mile**
Commuter rail **108 g CO_2 per passenger km/6.1 oz CO_2 per passenger mile**
Cycling **17 g CO_2e per km/ 1 oz per mile**

AVIATION
0.18
(0.2)

Long flight **138 g CO_2e per km/ 7 ¾ oz per mile**
Short flight **120 g CO_2e per km/ 6 ¾ oz per mile**

HEALTH AND HYGIENE
(Including bathing, showering, washing, and health services)
0.36
(0.4)

5-minute hot shower **1.5 kg (3.5 lbs) CO_2**
Bath **4 kg (9 lbs) CO_2e per day**
Laundry washed at 40°C (104°F) and tumble dried **2.5 kg (5 ½ lbs) CO_2**

EDUCATION
(Schools, and books and newspapers)
0.13
(0.14)
Daily newspaper, recycled, **400 g (14 oz) CO_2e**

GOVERNMENT AND DEFENCE
0.08
(0.09)

What can I do?

- **Go online and use a carbon footprint calculator** to see where your emissions are coming from.
- **Identify where savings can be made**. Once you have calculated your carbon footprint, you can work out a plan to cut carbon while saving money.
- **Consider what you eat.** Food is a major part of a person's total carbon footprint in most western countries, especially where there is a substantial element of meat and dairy produce.

Renewable revolution

Renewable energy sources are rapidly expanding, especially solar and wind power technologies. These and other clean energy sources will be vital for meeting rising demand while simultaneously combating climate change.

The advantage of renewable energy is that it can be replenished indefinitely, without depleting finite resources such as fossil fuels. Renewables can be used to provide electricity and heat, and to make transport fuel.

At present, electricity-generating wind and solar power technologies form the biggest, fastest-growing areas of the renewables sector.

Biogas (the same as fossil natural gas, but made from organic matter such as food waste) and wood can be used for heating as well as electricity. Liquid biofuels provide a renewable alternative to fossil-derived diesel and petrol.

Renewable energy can help address many environmental issues, as well as creating jobs and driving technological development.

Growth of renewable energy

Renewable energy is the fastest-growing source of power worldwide, with solar and wind power becoming ever more cost-effective. Several countries have undertaken significant investment in renewable power, and renewables accounted for 70 per cent of net additions to global power capacity in 2017. Projections for the future vary; it has been estimated that by 2030 renewables will overtake coal, and that by 2040 their usage will be equal to coal and natural gas combined.

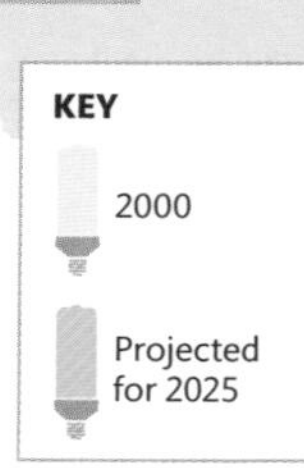

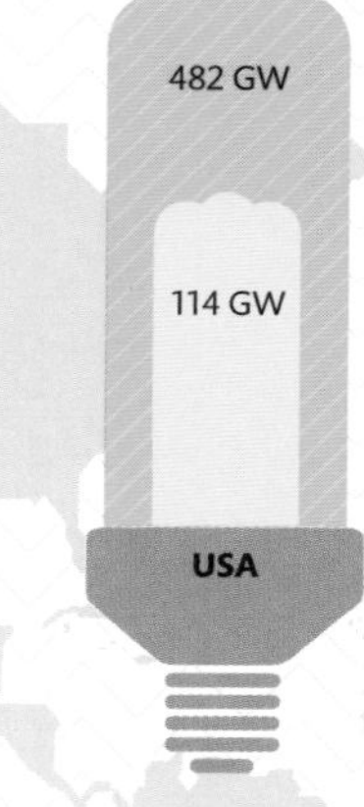

ELECTRICITY STORAGE

Because the wind doesn't blow all the time, and the sun sets at night, energy-storage technologies are vital to some sources of renewable energy. There are several options already available, and further research and development is underway, alongside investigation into new technologies. With sufficient storage capacity, the world can move to 100 per cent renewable energy, including the electrification of transport and heating, thereby enabling the potential to end the use of gas for warming homes and petrol and diesel in vehicles. Storage is thus vital in meeting the goal for net zero greenhouse gas emissions.

BATTERIES
Different battery technologies can store power. They require materials such as copper, lithium, and lead to make.

PUMPED WATER
Using wind-generated energy, water can be pumped uphill into reservoirs to be released later to drive hydro-power.

HYDROGEN
Hydrogen can be split from water using clean electricity, stored and then used later to generate electricity.

COMPRESSED AIR
Excess electricity can be used to pump compressed air underground. When released it can drive turbines.

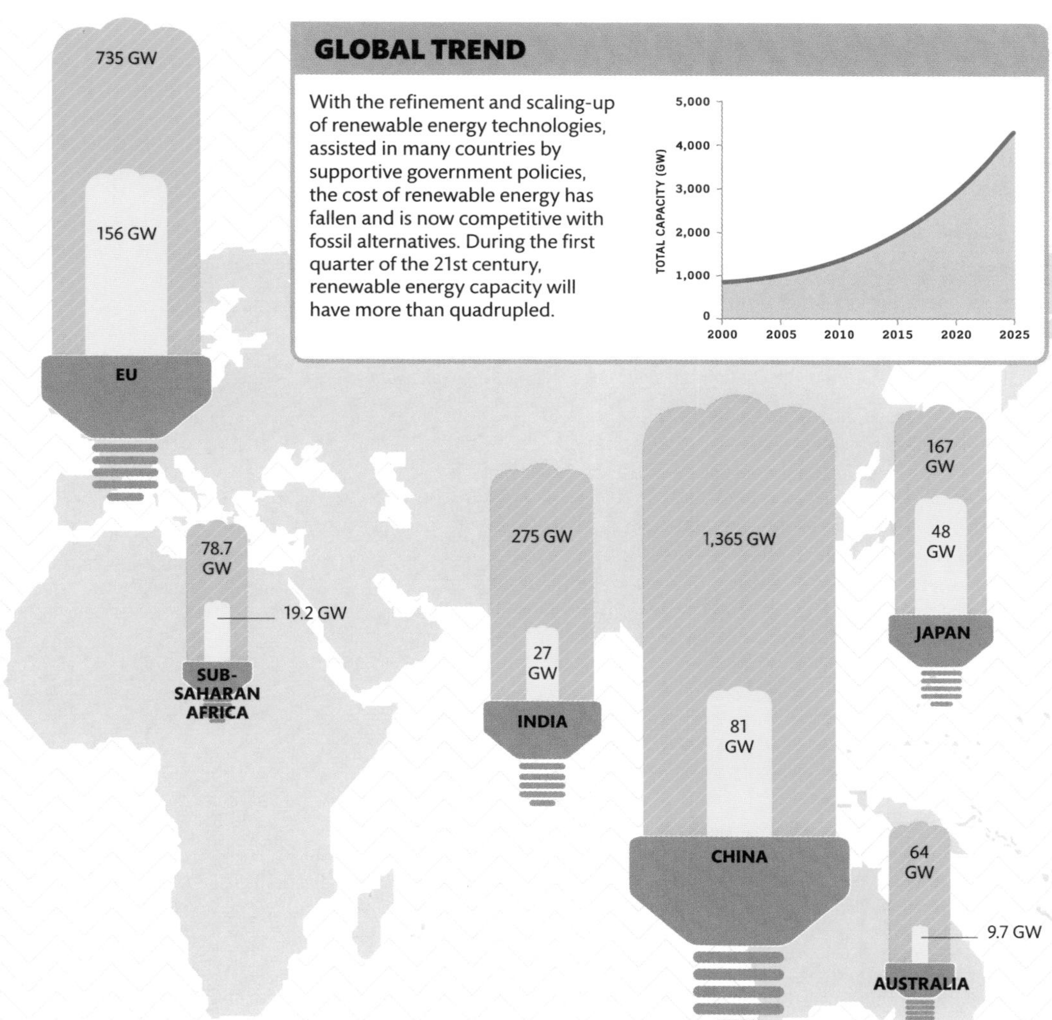

GLOBAL TREND

With the refinement and scaling-up of renewable energy technologies, assisted in many countries by supportive government policies, the cost of renewable energy has fallen and is now competitive with fossil alternatives. During the first quarter of the 21st century, renewable energy capacity will have more than quadrupled.

Measuring sustainable power
The graphic shows the growing scale of renewable energy capacity across the world expressed in gigawatts (GW) generated.

Renewables accounted for almost **22% of global electricity generation** in 2013 – up 5% from 2012

How solar energy works

The Sun is the ultimate source of energy for nearly all life on Earth. With the right technology, our home star could also be the main power station providing the energy needed to run the human world.

Solar photovoltaic (PV) panels

These use semi-conducting layers, usually silicon, to capture solar energy. Light hits a panel, creating an electric field across its layers that creates a current by separating positive and negative charges. The stronger the sunlight, the more electricity produced.

Solar powerhouse

The Sun emits a vast quantity of energy. The solar energy hitting Earth is sufficient to power around 4 trillion 100-watt light bulbs. The recent refinement of solar energy technologies and the rapid growth in their use lead many experts to believe that, by 2050, solar will be the world's principal energy source.

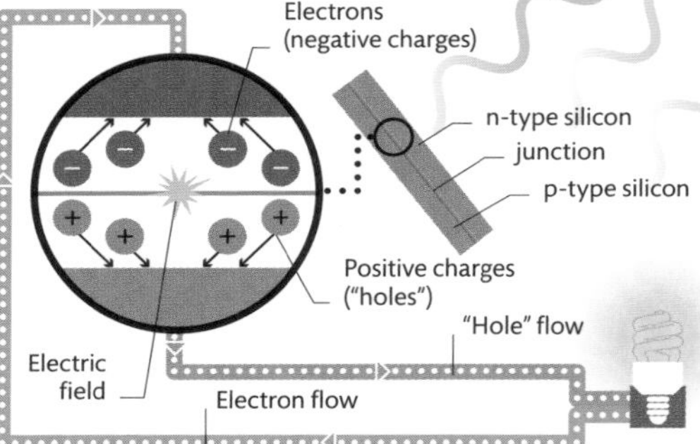

Concentrating solar power (CSP)

CSP linear concentrators, dish engines, and power towers (illustrated) use mirrors to focus the Sun's heat onto vessels carrying liquid, such as molten salt, that is heated to boil water. This creates steam, which drives electricity-generating turbines. Heat-storage facilities enable this form of solar power to generate electricity at night.

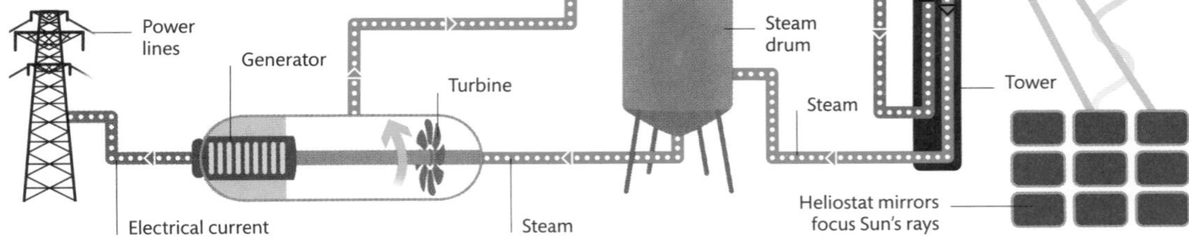

We have always relied on solar energy. For example, the horses that were once a main means of transporting people and goods were fed on grass and grains grown with sunlight. Today, however, new technologies are allowing us to make more use of solar energy, by converting heat or light from the Sun into more usable forms of energy, such as electricity and hot water. Solar technologies have pros and cons, but in all cases they also offer massive potential. Increased use and refinement is already leading to falling costs and potentially vast growth in the years ahead.

As the world struggles to cut climate-changing emissions, solar energy technologies are positioned to take over from fossil fuels.

Passive solar energy

Windows positioned to receive maximum natural light can cut the electricity needed to power light bulbs. Solar heating of interior surfaces reduces the need for internal heating, especially if a building is carefully insulated.

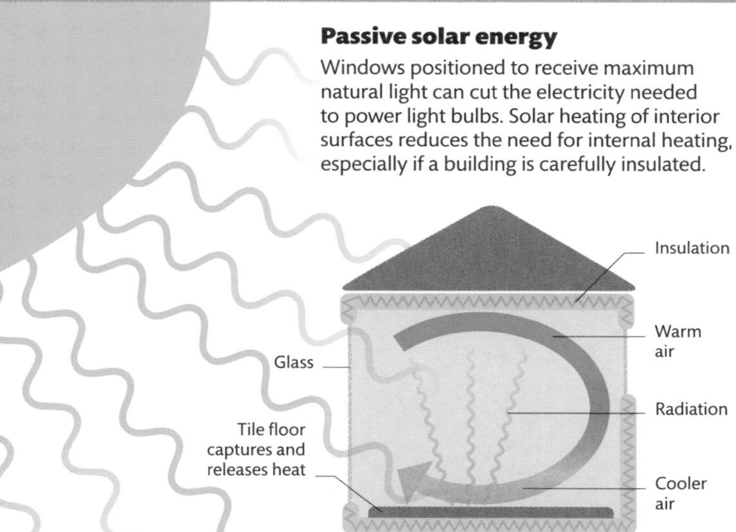

GLOBAL HOTSPOTS

Solar energy technologies can work almost anywhere there is a good amount of daylight. They are most effective, however, in regions where there is consistently strong sunshine and little cloud. Many desert areas and other sunny parts of the world have the potential to produce huge quantities of electricity by using existing solar energy technologies, such as solar PV and concentrating solar power. These include the southwestern USA, western South America, Africa, the Middle East, South Asia, and Australia.

Solar water heating

Solar water heating systems use solar panels called collectors to accumulate warmth from the Sun and use it to heat water stored in a hot water cylinder. A backup boiler or immersion heater can be used to heat the water further, especially at high latitudes during winter months.

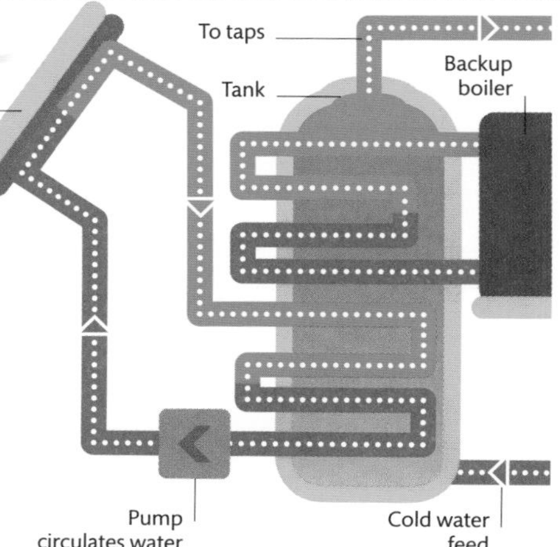

1 hour of sunlight hitting the Earth is roughly equal to the **planet's annual energy consumption**

Wind power

During recent decades, the use of wind-generated electricity has expanded rapidly in some parts of the world. Some countries, such as Denmark, now rely heavily on wind to supply much of their power.

In ancient times, wind energy was used to propel boats along the River Nile, pump water, and grind grain. By about 1000 CE, it was used to drain large areas of the Rhine delta. Wind was first harnessed to generate electricity in Glasgow, Scotland, in 1887. In 1941, the world's first megawatt turbine was connected to the power grid in Vermont, USA, followed by the first multi-turbine wind farm in New Hampshire in 1980, and the first offshore installation in Denmark in 1991. Since these pioneering wind farms were constructed, the technology has improved – and rapid growth has followed.

Who's generating the most?

A number of countries have adopted policies to encourage the installation of wind-powered electricity generation. In 2019 China had the world's biggest wind-power sector, followed by the USA, although recently it has added far less new capacity than China. Germany comes third, with 11 per cent of the world's wind power. Other major wind energy producers include India, Spain, the UK, Canada, France, and Brazil. Some countries, such as the UK, are installing huge new turbines sited offshore.

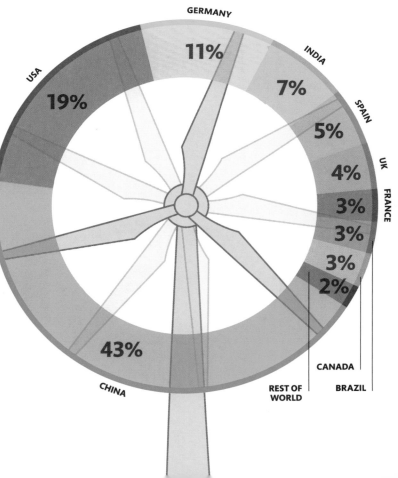

GERMANY 11%
INDIA 7%
SPAIN 5%
UK 4%
FRANCE 3%
3%
3%
2%
CANADA
BRAZIL
REST OF WORLD
CHINA 43%
USA 19%

Offshore wind power
Stronger ocean winds provide more electricity than wind farms on land, but offshore set-up costs are higher.

Blades rotate
As sufficient winds blow, the passing air pressure causes the turbine blades to revolve.

2

Gears spin generator
The blades turn a shaft connected to a gearbox, which increases the rotational energy produced.

How does wind power work?

Conventional generators use steam to drive turbines. With wind power, the process is powered by air instead of fuels such as coal or gas. The propeller-like blades are attached to a rotor that is connected to a main shaft, which in turn spins the generator. The whole assembly is mounted on a tower in order to take advantage of steadier and less turbulent winds.

"The future is **green energy,** sustainability, **renewable energy"**

ARNOLD SCHWARZENEGGER, FORMER GOVERNOR OF CALIFORNIA

3

Power out
The rotational energy is converted into electricity by a generator.

4

Transformation
A transformer converts the electricity into the correct voltage for distribution.

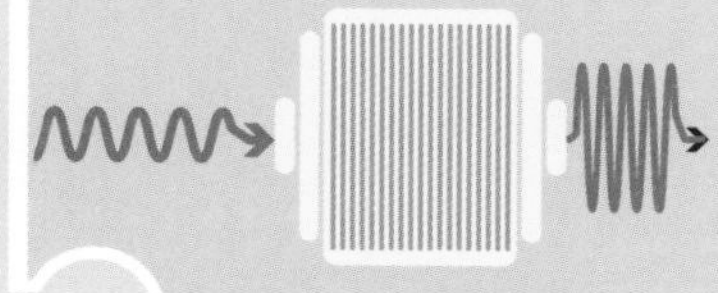

WIND-GENERATED ENERGY: PROS AND CONS

Pros

- It's clean, green, and pollution-free. Wind turbines create no emissions.
- It's renewable. Winds originate from solar energy, so promise an endless supply.
- Prices have decreased by 80 per cent since 1980 and will fall further. Operating costs are low.
- Potential for rapid growth.
- Technology improving to produce more power more quietly.

Cons

- Turbines typically operate at 30 per cent capacity.
- Can cause hazards to birds and bats. Soil erosion can be a problem during installation.
- Still more expensive than coal or gas-generated power in some countries.
- Can cause visual changes to landscapes.
- Viable only in areas of land and sea with sufficient steady wind.

5

Distribution
Electricity is sent throughout the country via a national grid network of cables.

Tidal and wave power

The seas hold vast quantities of energy that we are only beginning to convert into electricity through wave and tidal power systems. Like wind and solar technologies, they can produce pollution-free power.

Turning the tides

Tidal and wave energy technologies are becoming commercially viable power sources. The technology is advancing rapidly and has huge potential in coming decades. Wave farms and tidal energy systems harness the enormous power of the seas to generate power, and their global capacity could exceed that of about 120 nuclear reactors. Countries with the most potential for these reliable, renewable energy sources include France, the UK, Canada, Chile, China, Japan, Korea, Australia, and New Zealand.

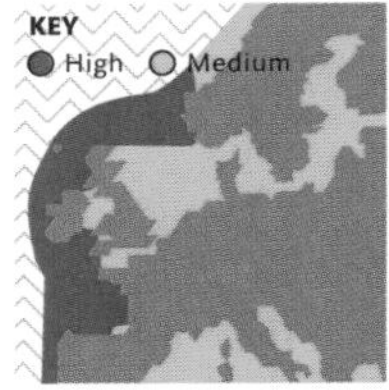
KEY
High
Medium

BEST SITES FOR WAVE POWER, EUROPE

Making waves

Europe's best areas for wave farms are along the western Atlantic coast, where strong, persistent winds create lots of large waves.

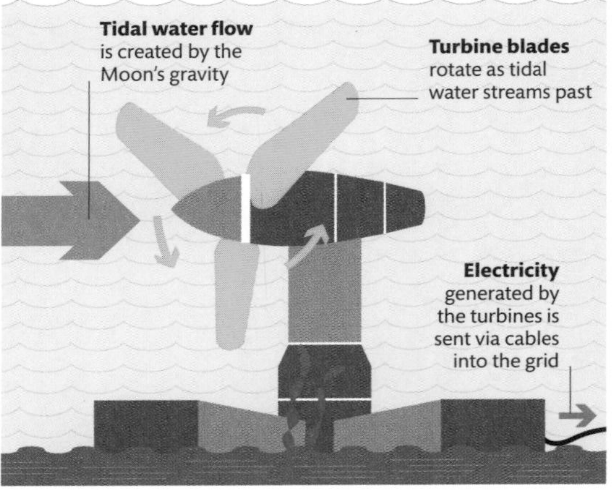
Tidal water flow is created by the Moon's gravity
Turbine blades rotate as tidal water streams past
Electricity generated by the turbines is sent via cables into the grid

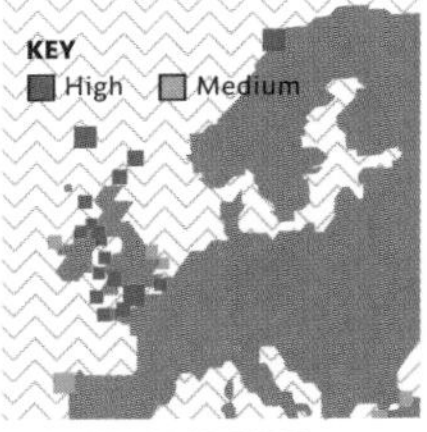
KEY
High
Medium

BEST SITES FOR TIDAL STREAM POWER, EUROPE

Streaming tides

Around the UK in particular, headlands, inlets, and channels funnel and increase the speed of tidal stream currents – ideal for tidal energy.

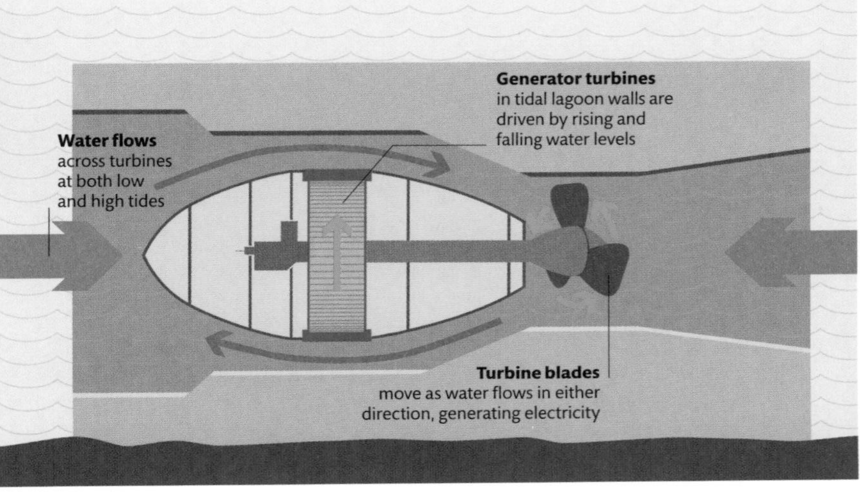
Generator turbines in tidal lagoon walls are driven by rising and falling water levels
Water flows across turbines at both low and high tides
Turbine blades move as water flows in either direction, generating electricity

Tidal and wave technologies harness the movement of tides and waves to drive electricity-generating turbines. As well as cutting carbon dioxide emissions, these technologies could offer energy security and create jobs.

Electricity produced by wave and tidal power is currently priced higher than that generated by fossil fuels – partly because fossil fuels are burned (and valued) without taking the costs of the climate change they cause into account.

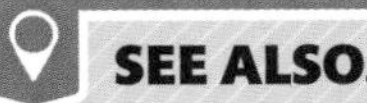

SEE ALSO...

- **Surge in demand** pp42–43
- **Renewable revolution** pp46–47
- **Energy conundrum** pp54–55

80%
The potential **kinetic energy** from waves that can be **converted into electricity**

CASE STUDY

Swansea Tidal Lagoon

- Swansea Bay in South Wales is located on the Bristol Channel. Because this area of UK coastline has the second-highest tidal range in the world, it provides an ideal location for a tidal lagoon.
- Sixteen underwater turbines are planned to be embedded in a breakwater wall extending 3 km (2 miles) out to sea.
- The tidal lagoon's proposed power station will generate clean, predictable power for more than 155,000 homes for at least 120 years.

HARNESSING SURFACE WAVES

One of the most promising designs for capturing surface wave power is the wave attenuator. Waves are best along western coastlines due to stronger, more consistent winds, so hotspots for this technology include the Pacific USA, the UK, France, Portugal, New Zealand, and southern Africa.

Tethered chain
Tied to the sea bed, the chain is parallel to wave direction

Hinged joints
Motion of hinges pressurizes hydraulic pistons, called rams

Electricity
Pressurized rams turn interior turbines, creating electricity

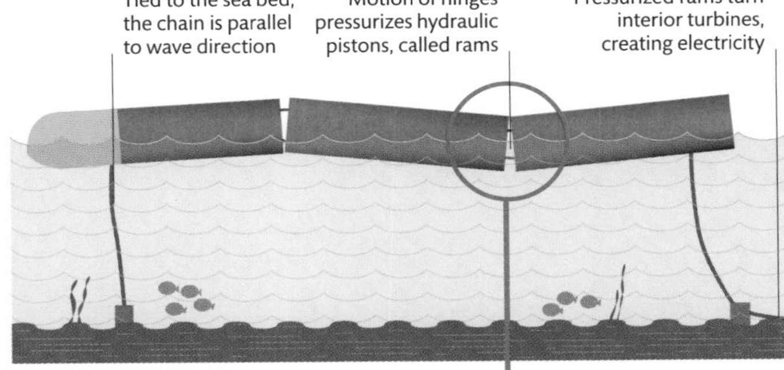

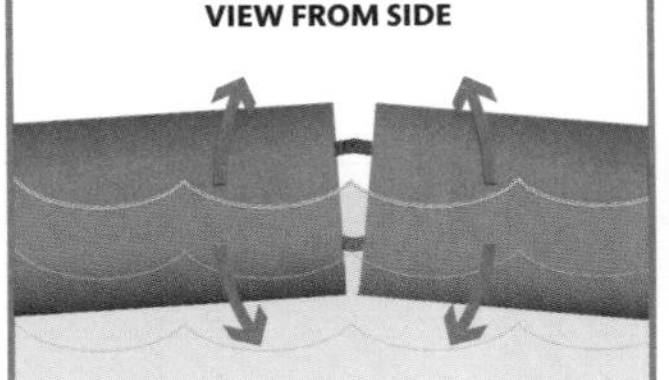

Vertical bends
The semi-submerged sections of a wave attenuator move vertically up and down with wave motion, flexing at their hinges.

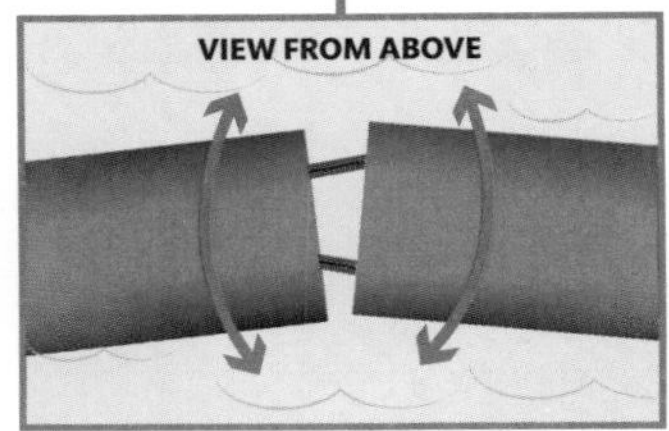

Oscillating motion
As well as being forced up and down, attenuator hinges also allow sections to "yaw", capturing wave energy from rotational movement.

Energy conundrum

Pros and cons accompany all of our energy choices. As rising demand leads to deepening tensions between competing priorities, it is vital that we see the full picture in order to make informed decisions.

Many different types of existing and emerging technologies play vital roles in meeting our energy needs. Parallel technologies will shape future choices: for example, carbon capture and storage in the case of coal and natural gas, and energy storage in relation to some renewables.

Our approach to energy must address the issues of security, affordability, and environmental impact – three goals that often pull in different directions. For example, coal provides cheap, secure power but causes high carbon dioxide emissions and air pollution.

Energy policy is a highly political issue. Decision-making often favours short-term cost and security objectives at the expense of environmental concerns. Factors like these make it all the more challenging to put most rational – and globally beneficial – long-term choices in place.

What are our options?

The simple comparison below is based on situations as they broadly prevail today. While circumstances for some technologies are highly variable - such as the potential for renewable energy sources in certain locations - some overall conclusions about each particular source of energy can be provided. Policy makers must decide which yield the best range of long-term results.

Coal

9

The single largest source of electricity worldwide, with recent massive growth in demand from fast-growing countries, including China and India.

- Abundant supply fuels cheap electricity.
- High carbon emissions and local air pollution.

Oil

10

The world's main transport fuel.

- A major source of carbon dioxide and urban air pollution.
- Oil produced by hydraulic fracturing (fracking) and tar sands creates higher carbon emissions than conventional oil.

Natural gas

4

Flexible, abundant, and used for electricity, heating, and cooking.

- Produces about half as much carbon dioxide as coal.
- Conventional gas and that produced by hydraulic fracturing (fracking) raise different issues.

Nuclear

8

Produces low-carbon electricity but is expensive and complex.

- Major issues are linked with long-term radioactive waste management.
- Tensions persist over the link between nuclear power and nuclear weapons.

Hydropower

5

Relatively low-carbon power source but limited by number of suitable rivers.

- Can lead to major ecosystem and social impacts.
- Vulnerable to prolonged droughts that are already affecting some regions.

KEY TO SYMBOLS AND RATINGS

Cost Energy costs often dictate choice, and are especially important for those on low incomes

Tech ready? Some technologies are well established, while others are just coming on stream

Pollution and waste Some technologies are much cleaner than others

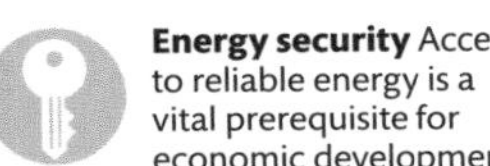

Energy security Access to reliable energy is a vital prerequisite for economic development

Land and ecosystems impact Energy supply can conflict with other resource and environmental goals

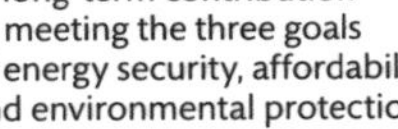

Overall rating the extent of long-term contribution to meeting the three goals of energy security, affordability and environmental protection.

1 Best

10 Worst

Strong case

Advantages

Drawbacks

Major concerns

EFFICIENCY: THE INVISIBLE "FUEL"

The most neglected fuel source is efficiency. Cars that are more fuel-efficient, lights that use less power, insulation, and smart building technologies all save energy without affecting comfort or convenience. Efficiency can also save money, making this an obvious priority when it comes to finding the best ways of reaching the three energy goals.

The global use of **energy efficient** measures could **save US$340-500 billion** a year by 2030

Liquid biofuel

Can replace fossil oil and bring carbon dioxide savings – for example, by using sugar cane to make ethanol.

- Can divert food supply from plates to fuel tanks.
- Can drive deforestation, leading to CO_2 emissions and biodiversity loss.

Biomass

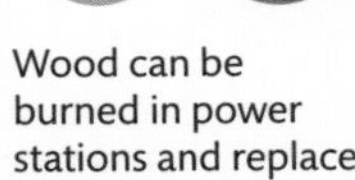

Wood can be burned in power stations and replace gas and coal.

- Renewable, but can lead to high carbon emissions and soil damage.
- Can drive deforestation.

Wind

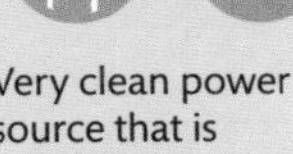

Very clean power source that is growing fast.

- Intermittent wind means other power sources are needed to meet constant demand, but energy-storage technologies are developing.
- Changes the appearance of landscapes.

Solar

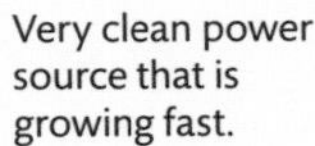

Very clean power source that is growing fast.

- Depends on daylight, so very large-scale use will rely on emerging storage technologies, such as large-capacity batteries.
- Its use is rapidly expanding worldwide.

Wave / Tidal

Very clean and potentially very significant power sources.

- Technologies are emerging, with first commercial power stations being installed.
- Relatively expensive. Needs government backing during start-up phase.

Escalating appetite

The rise of agriculture has transformed the face of planet Earth and shaped the course of human history. Pre-farming hunter-gatherer societies had a total population of a few million people, but today farming sustains more than seven billion people worldwide. The rise of high-productivity farming was a vital factor in the establishment of civilization and it enabled the continuing shift of people from rural areas to towns and cities. Sustaining the conditions that allow farming, including soil health and freshwater availability (pp78–79), are increasingly important challenges.

Grain production

Early farmers domesticated wild grasses to produce grains, including rice, wheat, and corn. Rich in carbohydrate and protein, easy to store, and fast growing even in quite poor soils (and, in the case of wheat, dry lands), grain became the backbone of agriculture. It remains so today, although new varieties, mechanization, pesticides, and fertilizers enable far larger yields than was possible even in the middle of the 20th century. Despite the rapidly growing population, the world has kept pace with the demand for an increased food supply and grain production has risen steadily since 1950.

Green Revolution research to find ways to increase crop yields begins in Mexico during the 1940s. Technologies such as fertilizers, pesticides, mechanization, and irrigation, spread worldwide during the 1950s and 60s.

1950 1955 1960 1965 1970 1975 1980

YEAR

THE RISE OF MEAT AND DAIRY

As people have become wealthier the consumption of meat and dairy products has risen dramatically. This has had negative effects both on the environment and on human health. Compared with vegetable-based diets, livestock-derived foods require more land and water to produce. Increased consumption of meat and dairy foods, both of which are high in protein and fat, raises the risk of heart disease, some cancers, and type 2 diabetes.

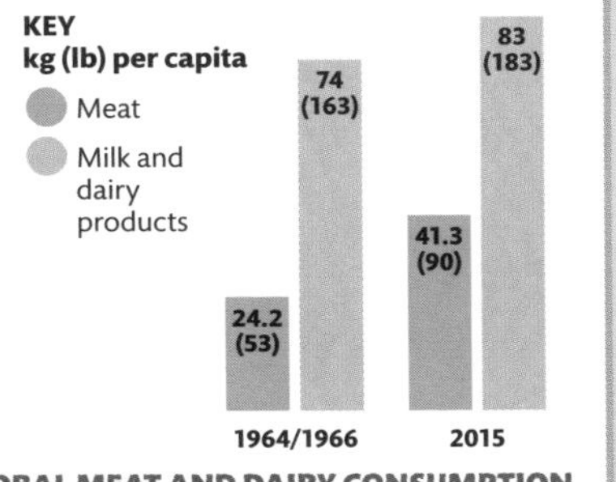

GLOBAL MEAT AND DAIRY CONSUMPTION

1997
First genetically modified corn is produced

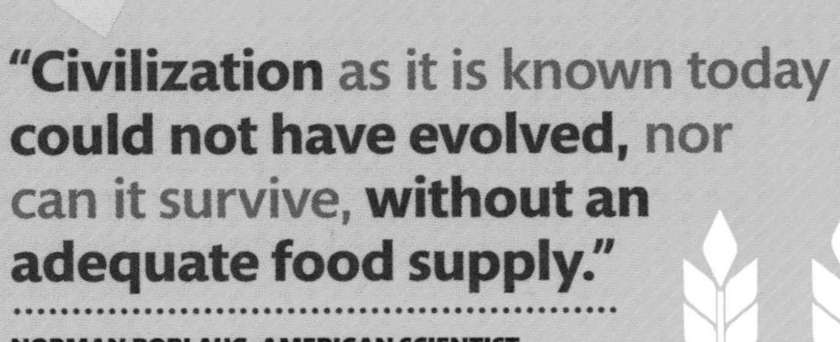

"Civilization as it is known today could not have evolved, nor can it survive, without an adequate food supply."

NORMAN BORLAUG, AMERICAN SCIENTIST AND "FATHER" OF THE GREEN REVOLUTION

Global grain production
In 2018, China produced one fifth of global grain output. In that year the world's total cereal production was estimated at 2,962 million tonnes (3,265 tons), mostly comprised of corn, rice, and wheat.

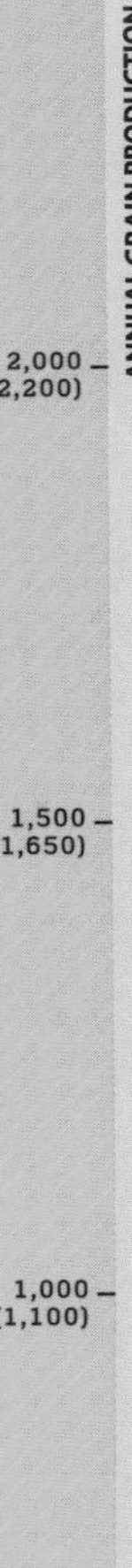

Farmed planet

About one third of the world's land is now used for farming. Only about one quarter of that third is used to produce crops, however, with the rest used to rear animals.

Most of the world's land is desert, covered with ice, or supports forests and grasslands, much of which is unsuitable for farming. Where conditions permit, there has been a steady expansion of agriculture, although the total land area with suitable soils and sufficient water for crop production is, in a global context, limited. Rising demand for food is leading to the continuing expansion of farming into the remaining unconverted areas, where there is suitable soil and sufficient water. The consequences of this include deforestation, declining wildlife, increased greenhouse gas emissions, declining water quality, and widespread soil damage (see pp68–69).

Crops versus meat

About three quarters of the world's land that is producing food is devoted to rearing livestock to supply meat and dairy products. The remaining land is used to produce grains, fruits, and vegetables. Consumption of livestock products has increased in line with the rise in the number of middle-class consumers. This is set to continue, as the major emerging economies shift their dietary preferences. Although only a fraction of farmland is cultivated for grain and vegetable production, a high proportion of the crops produced are fed to livestock. Grasslands, sparsely wooded country, and barren lands are also partly grazed by domesticated animals.

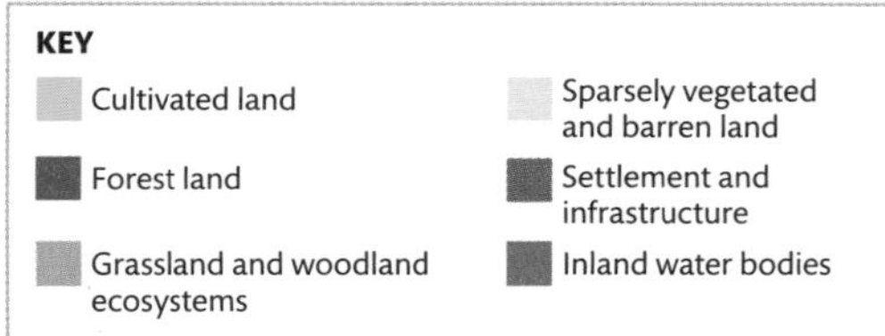

0 10 20 30 40 50

WORLD REGION

Large areas of boreal forest and tundra

NORTH AMERICA

Vast areas of tropical forests remain despite conversion to crops and clearance for ranching

SOUTH AMERICA

SUB-SAHARAN AFRICA

Large-scale conversion of good soils for rice production has enabled high population density to be sustained

SOUTH ASIA

A high proportion of desert and barren land and low rainfall limits the extent of crop production

EAST ASIA

A high proportion of tropical rainforest remains, although it is under pressure from agricultural expansion

SOUTHEAST ASIA

Extensive fertile soils and abundant rains permit large-scale crop production

CENTRAL AND WESTERN EUROPE

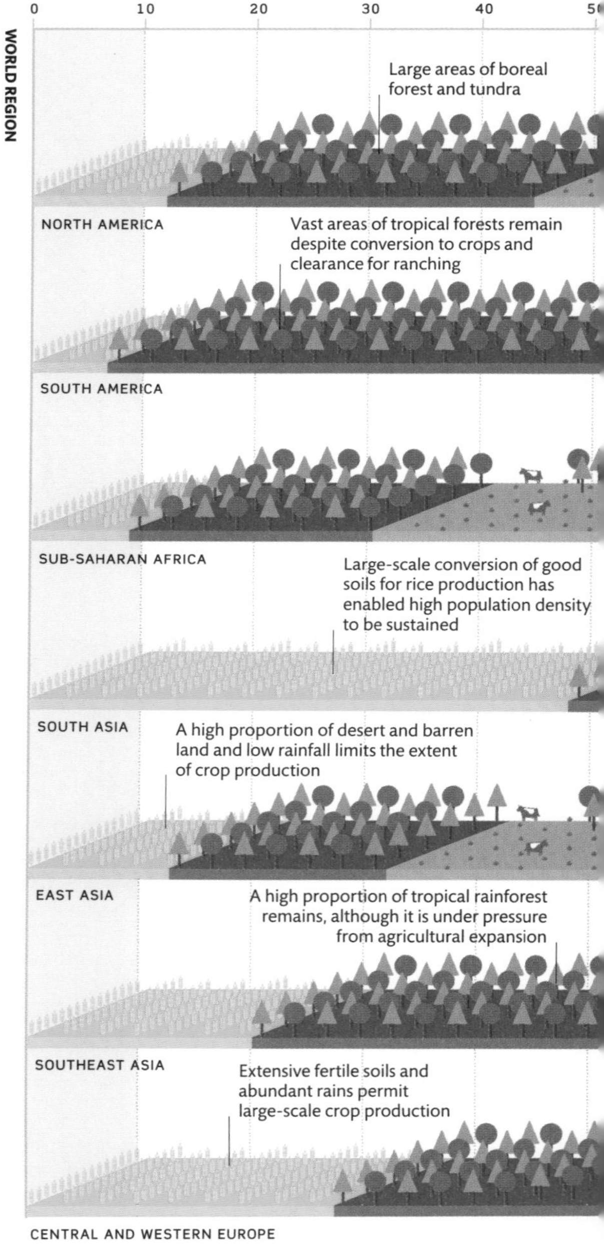

LAND USE AS PERCENTAGE OF WORLD REGION

70 80 90 100

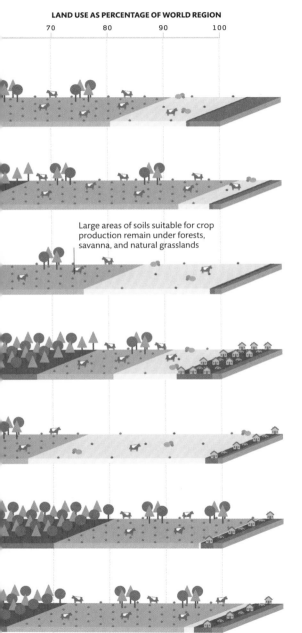

Large areas of soils suitable for crop production remain under forests, savanna, and natural grasslands

Change over time

The rise of agriculture during the last two centuries has been dramatic. In 1800, most of the farmed land was in Europe and parts of Asia. Today, it has expanded across those continents and transformed the face of North and South America, and much of Africa and Australia, where natural vegetation has been cleared to make way for crops and livestock.

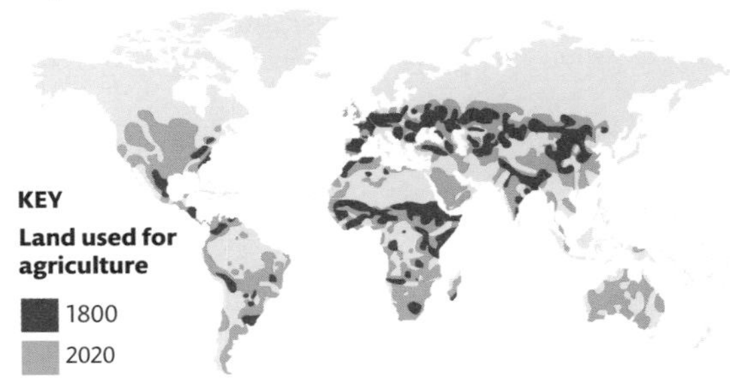

KEY

Land used for agriculture

1800

2020

CEREAL USES

Each year, the world produces around 2.8 billion tonnes (3.1 billion tons) of cereals. Rice and wheat are mostly consumed by people, but most corn is fed to livestock. Feeding crops to animals, which are then consumed by people, uses more land, water, and fossil fuels than people eating crops directly.

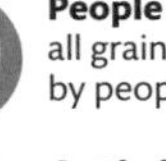

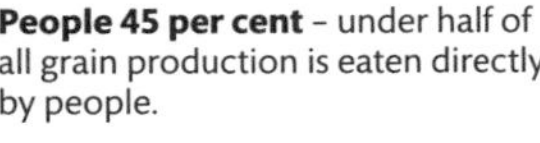

People 45 per cent – under half of all grain production is eaten directly by people.

Cattle feed 35 per cent – grains such as corn are used to feed pigs, cattle, and chickens.

Other uses 20 per cent – some grains have non-food uses, and are used for biofuels and industrial materials.

TOTAL LAND
13,003 million hectares (32,131 million acres)

LAND USED FOR AGRICULTURE
4,889 million hectares (12,080 million acres)

TOTAL AGRICULTURAL LAND

Fertilizer boom

The dramatic increase in food production achieved during recent decades has rested in large part on a corresponding increase in the use of fertilizers. However, this success has brought major challenges.

The plants that sustain all people and animals need soil nutrients – including nitrogen, phosphorus, and potassium – to grow. These are depleted by agriculture and need to be replaced. For millennia, farmers used nutrients recycled from wastes, such as manure. Industrial farming is sustained by the input of fertilizer from other sources, which has had a major environmental impact.

Improving yield

The invention of the Haber-Bosch process in the first half of the 20th century enabled nitrogen fertilizer to be made using natural gas and nitrogen from the atmosphere. Large-scale fertilizer application allowed farmers to produce more food from the same land, thereby keeping pace with increasing demand. Between 1950 and 1990 world food production almost tripled, while farmland only increased by 10 per cent.

Rise of fertilizer use

In the aftermath of the Second World War, chemical factories began to produce nitrogen fertilizer. New sources of rock phosphate were identified and the availability of phosphorus increased. With encouragement in some countries from government subsidies, the use of fertilizer grew rapidly, especially during the "Green Revolution" from the late 1940s to 1970.

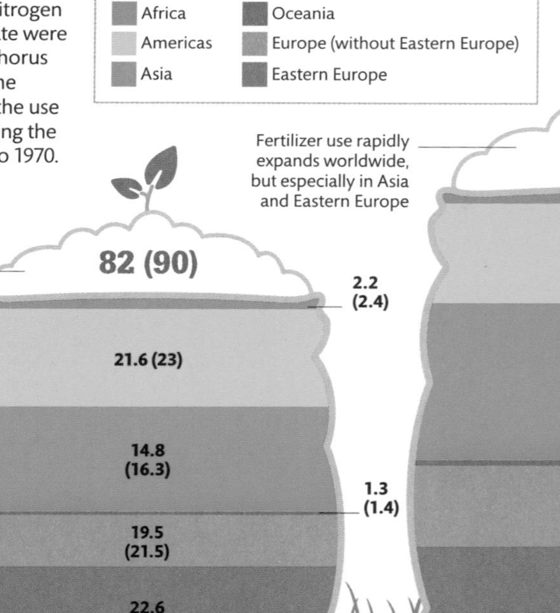

EFFECTS OF NITRATE FERTILIZERS

The main reason for the increased concentration of nitrous oxide in the atmosphere is the application of nitrogen fertilizers. These have a number of harmful effects on the environment and human health.

- Nitrous oxide is the third most important greenhouse gas causing climate change.
- Nitrogen fertilizers are partly responsible for the depletion of the ozone layer.
- Nitrogen (and phosphate) can cause ecological changes, especially in aquatic and marine environments, harming fish and other wildlife (see pp154–55).
- Fertilizer enrichment causes changes to ecosystems on land, thereby enabling more aggressive plants to displace more fragile ones.
- Nitrates building up in the environment can get into drinking water and present threats to human health. These include risk of "blue baby syndrome", various cancers, and thyroid conditions.

100% **increase in fixed nitrogen** on planet Earth over the last century due to **human activities**

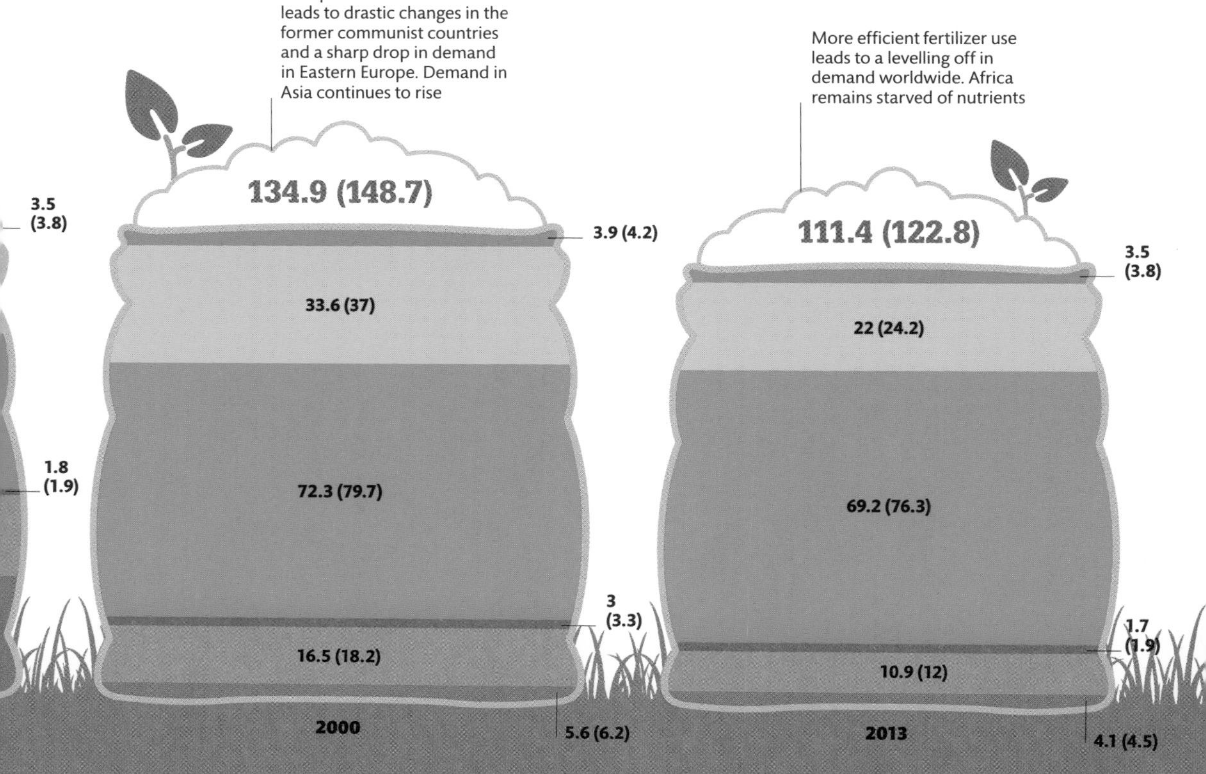

Pest control challenge

Weeds, fungi, microbes, and insects assault food crops, reducing yields and spoiling food. We have fought back with chemicals, but in the process have caused damage to wildlife.

For millennia farmers grew crops without chemical pesticides. In the decades after the Second World War, toxic compounds came into widespread and ever-increasing use, emerging as a key factor in the rapid expansion in food production. But the damage caused to wildlife has been considerable. The effects are wide ranging and include the loss of food plants for insects and a decline in the food supply for insect-eating birds. The populations of beneficial animals are affected too, including pollinators. Some pesticides accumulate in food chains, causing populations of top predators to decline (see pp86–87). At the same time, pests have developed resistance to pesticides.

How much pesticide is used

Pesticide use is rising almost everywhere, but countries vary widely in the quantity that they use. This is determined by the type of crops being grown, how valuable they are, and whether pest pressures are high. It also depends on the potency of the chemicals being applied, agricultural practices, and the stage of development the country has reached, with very poor countries unable to afford pesticides. Government policy and the extent to which pesticide companies have been successful in influencing policy also plays a part. In most cases, however, pesticide use could be reduced.

The amount of pesticides used internationally has risen **fifty-fold since 1950**

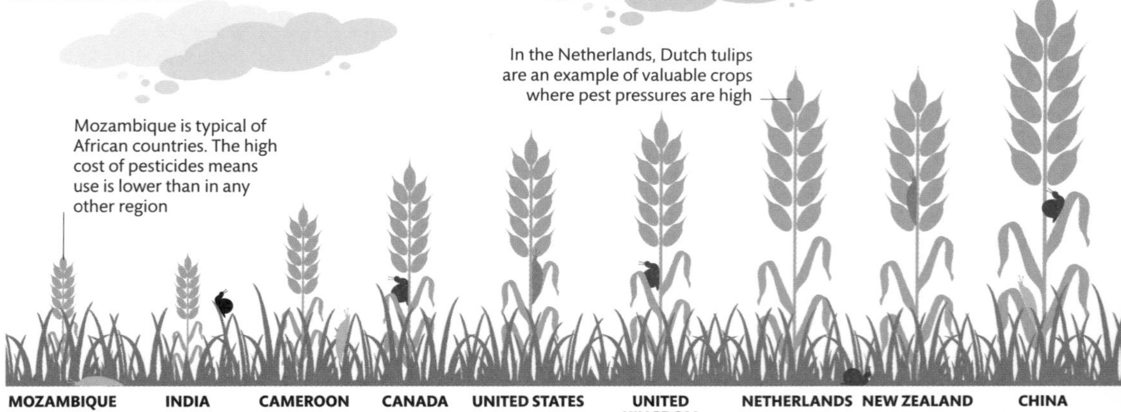

GLOBAL RISE IN PESTICIDE SALES

Global pesticide sales have been rising rapidly since the 1940s. Since 2000, sales have continued to increase, particularly in Asia, Latin America, and Eastern Europe. However, they have stagnated in the Middle East and Africa. Pesticide companies boost their sales by charging lower prices for older products or in poorer markets.

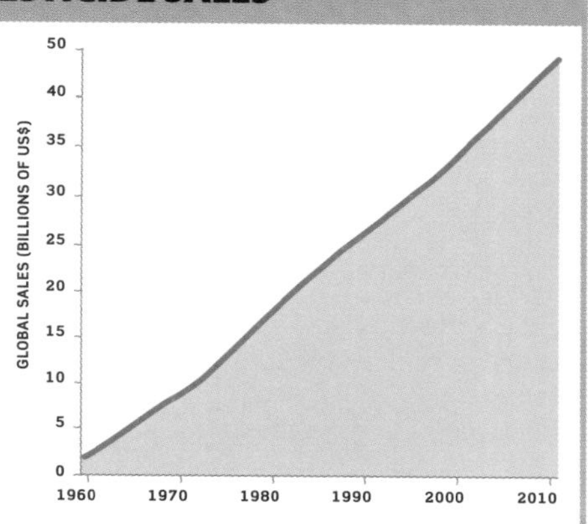

Pesticide application
Pesticides play an important role in growing rice in South and Southeast Asia. Spraying by hand is a common practice.

Threat to wildlife

Neonicotinoid pesticides are potent toxins that affect the nervous system of insects. Their use has affected many bird populations because insects comprise an important part of their diet. A study found that in areas with imidacloprid (a neonicotinoid pesticide) concentrations higher than 19.43 nanograms per litre, bird populations were in decline.

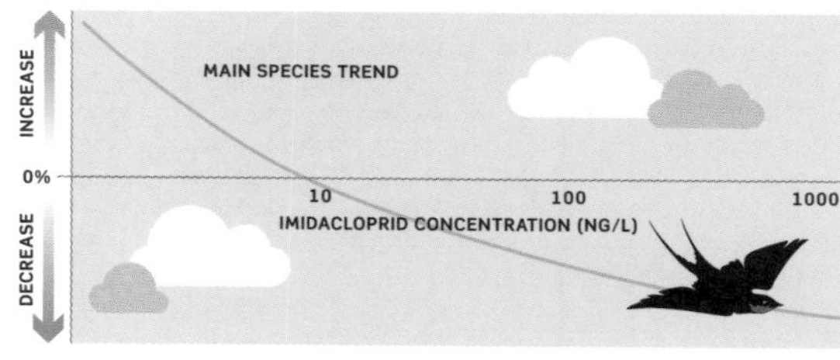

What can we do?

› **Governments, farmers, and chemical companies** can promote integrated pest management. This involves adopting strategies to enable food production with fewer chemicals, through growing a more diverse range of crops, and the use of crop rotations. Encouraging the recovery of bat and bird populations can improve natural pest management.

How food is wasted

The extent of food waste means that more than one quarter of the world's farmland is actually producing food for bins rather than people. As population and economic growth lead to rising demand, reducing food waste is an ever more important priority.

Worldwide, we waste about 1.3 billion tonnes (1.4 billion tons), or one third, of the food we produce every year. This in turn wastes water equivalent to the annual flow of Russia's massive Volga River. Food waste adds more than 3 billion tonnes (3.3 billion tons) of greenhouse gases to the atmosphere, not least because rotting food can create methane emissions that add to climate change. It wastes millions of tonnes of fertilizer and costs food producers US$750 billion annually. It also represents a missed opportunity to make sure everyone has access to food. The later a food item is spoiled in the journey from field to plate, the bigger the environmental impact because more resources will have been used up in getting it there.

Where is it lost?

Food waste occurs at every stage of the supply chain, from initial production right down to household consumption. In developing countries, 40 per cent of food losses occur at early stages of the process, and can be attributed to constraints in harvesting techniques and storage and cooling facilities. In developed countries, over 40 per cent of waste occurs at the retail stage, due to quality standards that over-emphasize appearance, or during the consumption stage when food is thrown away.

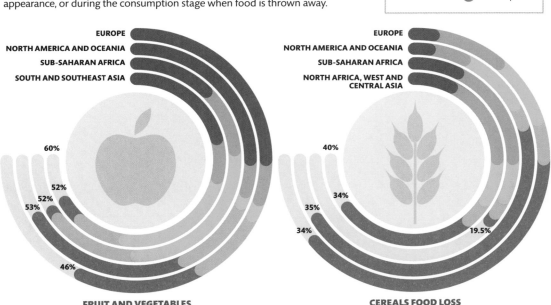

WHAT ARE WE WASTING?

All the major food groups are subject to substantial waste globally, but it is among the more fragile and perishable fruits, vegetables, roots, and tubers that the biggest proportion is lost. Meat waste is comparatively low, but the impact is bigger because calories from livestock farming come with a larger environmental footprint.

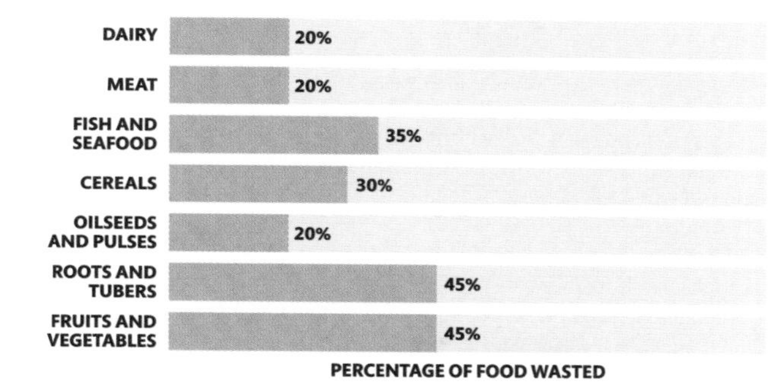

What can we do?

- **Reduce waste.** Avoid wasting food between farm and table.
- **Feed people in need.** Good food that would otherwise be wasted can sometimes be diverted to people in need.
- **Feed livestock.** Food unfit for human consumption can be fed to animals, such as pigs and chickens.
- **Compost and make renewable energy.** Badly spoiled food can be used to generate power via anaerobic digestion, while at the same time recovering nutrients that can be used as fertilizer.

EUROPE
NORTH AMERICA AND OCEANIA
SUB-SAHARAN AFRICA
LATIN AMERICA

30%
25%
22%
13%
20%

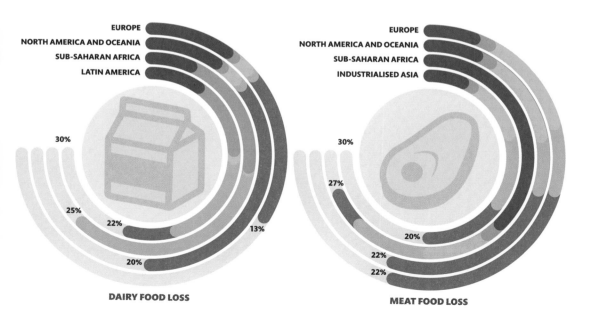

DAIRY FOOD LOSS

EUROPE
NORTH AMERICA AND OCEANIA
SUB-SAHARAN AFRICA
INDUSTRIALISED ASIA

30%
27%
20%
22%
22%

MEAT FOOD LOSS

Feeding the world

Across the world, hundreds of millions of people are hungry while hundreds of millions more are obese. This demonstrates how absolute levels of food production are not enough to ensure good nutrition.

In many rich and more developed countries, growing numbers of people are becoming overweight or obese, while in many developing countries a large proportion of people are undernourished. These outcomes are linked with various factors, including political and climatic conditions, and the proportion of their income people must spend on food. Despite an increase in food production during recent decades, poverty and hunger remain closely related. Inclusive economic growth is needed to improve the incomes and livelihoods of the poor, which would help to reduce hunger and malnutrition.

Where are the hungry?

In 2018, more than 820 million people worldwide were chronically undernourished. This was a significant increase compared with 2015. The situation was particularly severe in drought-sensitive countries, which are home to the poorest of the poor, who have limited financial assets and often live in rural areas. In sub-Saharan Africa, nearly one quarter of the population has insufficient food. India has the highest number of undernourished people, although they form a smaller percentage of the country's population compared with some others.

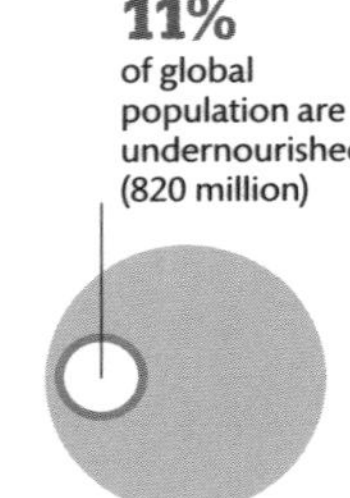

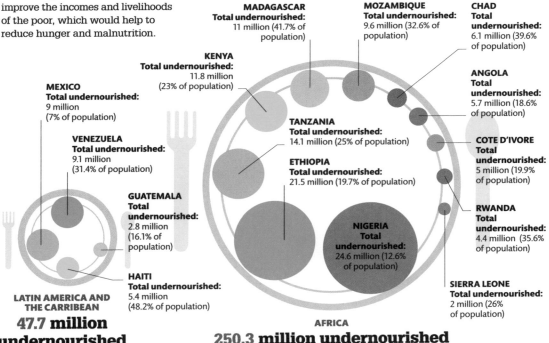

MADAGASCAR
Total undernourished:
11 million (41.7% of population)

MOZAMBIQUE
Total undernourished:
9.6 million (32.6% of population)

CHAD
Total undernourished:
6.1 million (39.6% of population)

KENYA
Total undernourished:
11.8 million (23% of population)

ANGOLA
Total undernourished:
5.7 million (18.6% of population)

MEXICO
Total undernourished:
9 million (7% of population)

TANZANIA
Total undernourished:
14.1 million (25% of population)

VENEZUELA
Total undernourished:
9.1 million (31.4% of population)

COTE D'IVORE
Total undernourished:
5 million (19.9% of population)

ETHIOPIA
Total undernourished:
21.5 million (19.7% of population)

GUATEMALA
Total undernourished:
2.8 million (16.1% of population)

RWANDA
Total undernourished:
4.4 million (35.6% of population)

NIGERIA
Total undernourished:
24.6 million (12.6% of population)

HAITI
Total undernourished:
5.4 million (48.2% of population)

SIERRA LEONE
Total undernourished:
2 million (26% of population)

LATIN AMERICA AND THE CARRIBEAN
47.7 million undernourished

AFRICA
250.3 million undernourished

THE COST OF FOOD

The price of food, in both absolute terms and relative to income, is an important determinant of hunger and obesity. In the USA, the average citizen spends a relatively small proportion of their large annual income on food. In India, the average person spends a large proportion of a tiny annual income on food.

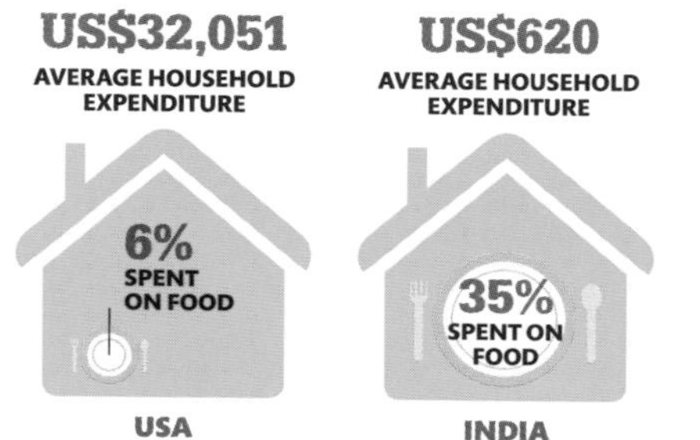

> "The **war against hunger** is truly **mankind's war of liberation**
>
> **JOHN F KENNEDY, 35TH US PRESIDENT, 1961**

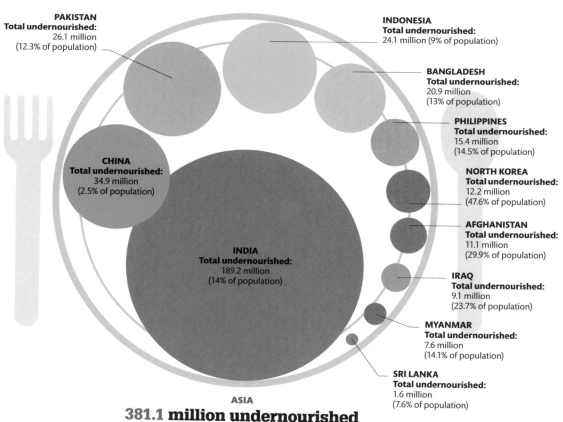

PAKISTAN
Total undernourished:
26.1 million
(12.3% of population)

INDONESIA
Total undernourished:
24.1 million (9% of population)

BANGLADESH
Total undernourished:
20.9 million
(13% of population)

PHILIPPINES
Total undernourished:
15.4 million
(14.5% of population)

CHINA
Total undernourished:
34.9 million
(2.5% of population)

NORTH KOREA
Total undernourished:
12.2 million
(47.6% of population)

AFGHANISTAN
Total undernourished:
11.1 million
(29.9% of population)

INDIA
Total undernourished:
189.2 million
(14% of population)

IRAQ
Total undernourished:
9.1 million
(23.7% of population)

MYANMAR
Total undernourished:
7.6 million
(14.1% of population)

SRI LANKA
Total undernourished:
1.6 million
(7.6% of population)

ASIA
381.1 million undernourished

Threats to food security

Nearly all food production depends on soil and freshwater. In both cases, environmental changes are leading to threats to food security. The challenge is global but becoming acute in many developing countries.

Every year 5–7 million hectares (12.4–17.3 million acres) of farming land are degraded with 25 billion tonnes (27.5 billion tons) of topsoil eroded by wind and water. Since settled agriculture began, the USA has lost about one third of its topsoil. Farming practices cause damage that can reduce the level of organic matter (decomposing plants and soil organisms). Soils with more organic matter hold more water, rendering growing plants more resilient to drought. In developing countries, soil damage and drought are prevalent. It is expected that later this century large portions of the world will experience extreme and, in some cases, unprecedented dryness.

Soil degradation

Soil damage is a widespread and worsening global problem. Human-induced soil degradation has already made many areas unsuitable for farming, especially in semi-arid parts of the world. Ploughing and excessive pressure from grazing animals can leave soils bare and vulnerable to removal by wind and rain. This is the cause of nearly all soil damage in North America. In South America, Europe, and Asia, deforestation is responsible for widespread soil damage. Relatively small areas of land have been damaged by industrial pollution.

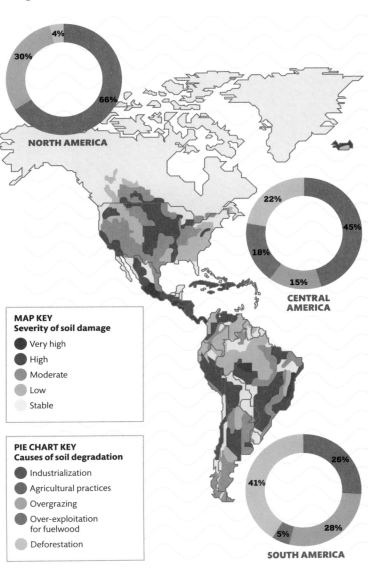

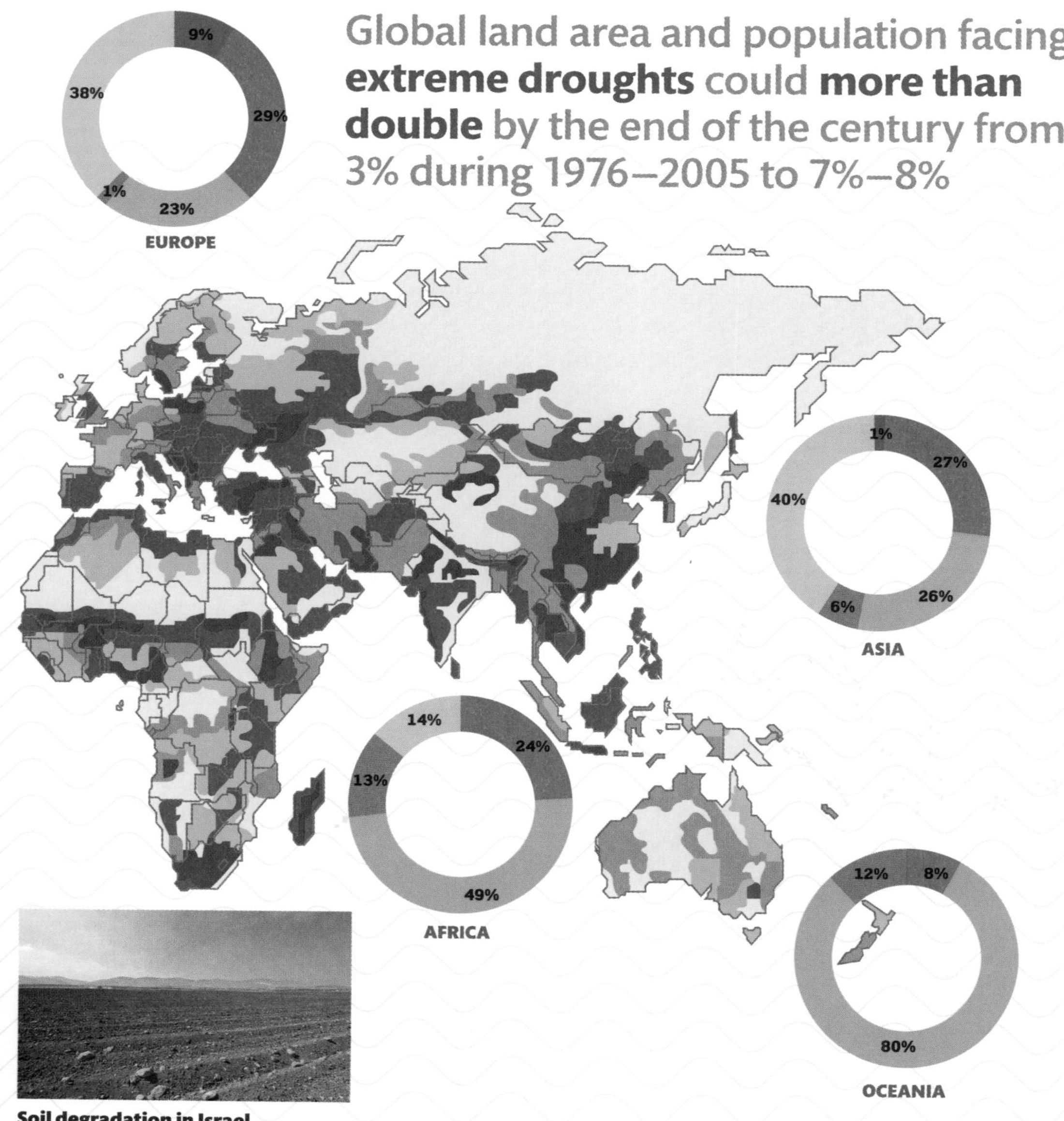

Soil degradation in Israel
Moderate to severe soil degradation globally affects an area of land larger than that of the USA and Mexico combined.

Thirsty world

Our need for freshwater has risen dramatically over the last century. As well as being required for drinking, washing, and agriculture, freshwater also helps power economic development. In the natural world, all land plants and animals rely on freshwater. Some ecosystems, such as tropical forests and wetlands, are dependent on regular replenishment of water. In recent years, several parts of the world have suffered from the effects of severe drought. The result has affected harvests and food prices, and increased the number of hungry people by millions.

Pressure on water supplies

Water covers 70 per cent of our planet, but less than three per cent of this is freshwater, and most of that is unavailable for our use (see pp72–73). Since 1900, population and economic growth have led to around a five-fold increase in water consumption. In some parts of the world, access to sufficient water is a serious constraint to development. Matters are made worse by the inefficient use of water in farming, industry, and homes, and through damage to ecosystems that help to replenish secure water supplies. Pressure on water resources can be expected to become more challenging still as the effects of climate change disrupt the water cycle, including the effect of more severe droughts, and areas already prone to water stress.

"A nation that **fails to plan** intelligently for the **development and protection of its precious waters** will be **condemned to wither...**"

LYNDON B JOHNSON, 36TH PRESIDENT OF THE UNITED STATES

1910
The invention of the Haber-Bosch process enables the industrial production of nitrogen fertilizer but leads to increased water demand

1952
USA passes Saline Water Act of 1952, marking the start of large-scale desalination of seawater

1900 1910 1920 1930 1940 1950

YEAR

WHERE WATER IS USED

More than half of the world's freshwater withdrawals occur in Asia, where most major irrigated lands are found. On average, however, water use per person is higher in richer countries, with people in the USA using about five times more than people in Bangladesh. In rich dry countries water stress is acute.

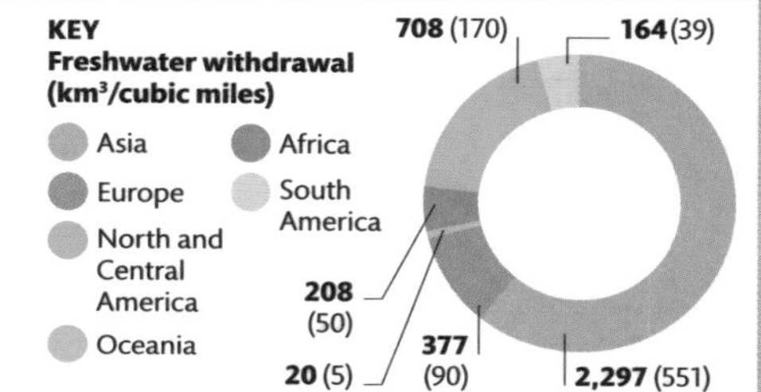

A series of record-breaking droughts and heat-waves lead to reduced crop production across the world.

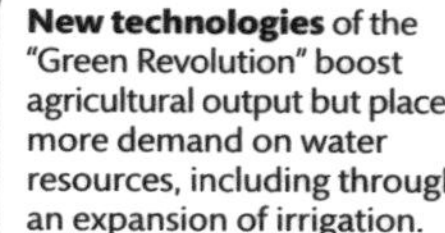
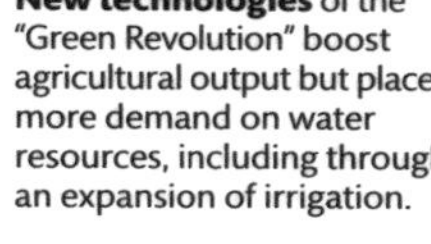

New technologies of the "Green Revolution" boost agricultural output but place more demand on water resources, including through an expansion of irrigation.

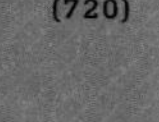

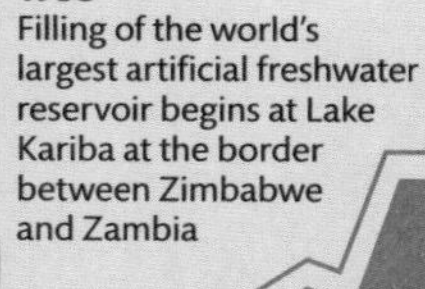

1958
Filling of the world's largest artificial freshwater reservoir begins at Lake Kariba at the border between Zimbabwe and Zambia

How freshwater is used

Although the proportion varies widely between countries, about 70 per cent of water withdrawn globally is used for agriculture. Agricultural, industrial, and domestic use are all projected to carry on increasing to 2025.

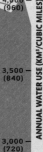

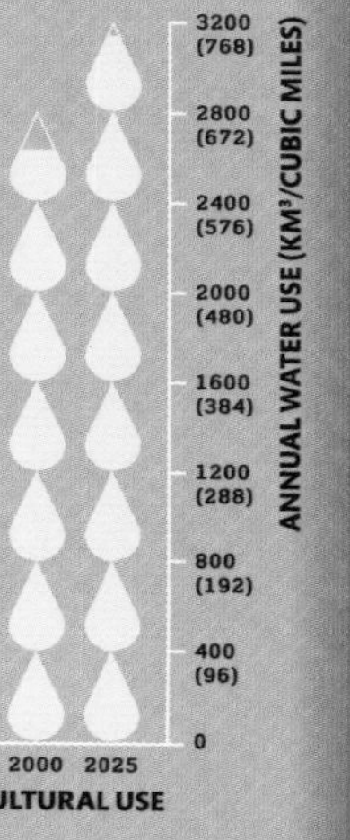

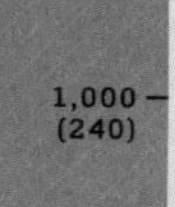
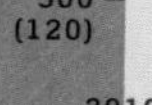

Freshwater scarcity

About 97.5 per cent of the world's water is in the oceans and salty. The rest is freshwater, but most of this is locked up in ice, with only about 0.3 per cent accessible for human use.

Freshwater is a surprisingly scarce resource. It is also unevenly spread, and in areas with low rainfall or high evaporation, scarcity can be a major problem. Water scarcity already affects 1.2 billion people across every continent. Another 1.6 billion people are affected by the challenges in extracting and transporting water. These numbers are rising, not least because water demand has been rising at more than twice the rate of population growth, causing the spread of long-term water scarcity to other parts of the world. Although too much of it is presently wasted, polluted, or used in unsustainable ways, there is still enough water on Earth to meet our needs. Making more rational use of water will be vital in the decades ahead.

SEE ALSO...

› **The population explosion** pp16–17

› **Escalating appetite** pp56–57

Earth's water resources

Almost all of Earth's 1.4 billion km³ (336 million cubic miles) of water is saltwater. Of the small portion that is fresh, more than two-thirds is locked up in ice caps, especially on Antarctica and Greenland. Nearly all of the remaining third is in the ground, and much of that is out of reach. This leaves only a tiny proportion as freshwater in the lakes and rivers from which we meet demand for drinking water as well as supplying farming and industry.

WATER
Life began in the oceans but spread to the land, where all animals and plants rely on freshwater

WATER-RICH NATIONS

Countries' economies rely on freshwater. Brazil's most populous region, São Paulo, suffered severe drought in the years 2014–17. With two-thirds of the country's power grid dependent on water reservoirs providing hydroelectric power, rationing is inevitable. Meanwhile, the continuing expansion of China's huge industrial output demands more and more freshwater.

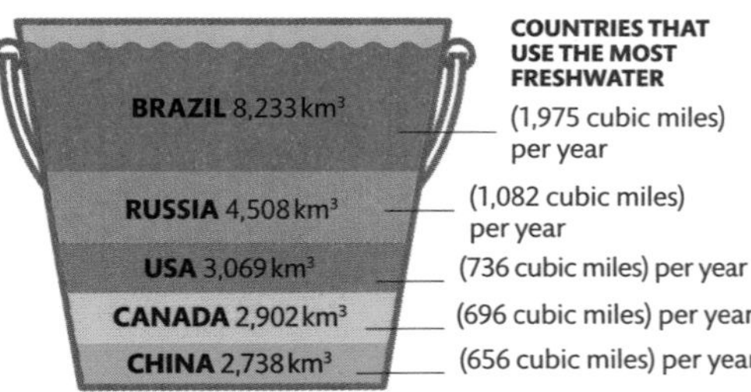

The surface of Earth is 71% water

LIQUID WATER
At just 0.3 per cent, a tiny proportion of the world's freshwater is in liquid form and readily accessible at the surface from rivers, lakes, and swamps

"When the **well is dry,** we learn the **worth of water."**

BENJAMIN FRANKLIN

ICE AND GLACIERS
The vast majority of freshwater is stored in glaciers, ice caps, and permanent snow cover in the mountains and polar regions of Earth

Total water on Earth 1.4 billion km³ (336 million cubic miles)

68.9% in glaciers and ice

GROUNDWATER
Of the freshwater on Earth, 30.8 per cent is groundwater. In some parts of the world, such as the USA and Arabia, fossil groundwater is being depleted to irrigate crops.

30.8% as groundwater

2.5% Freshwater

97.5% Saltwater

Freshwater sources

Ecosystems that store water include healthy soils, forests, and wetlands such as marshes and blanket bogs. Acidic peatlands in cool, wet climates also hold a lot of water. These environments are changing due to three main forces: global warming, which can change rainfall patterns and melt glaciers and ice caps; excessive water extraction to meet rising demand; and pollution, which contaminates an already limited water resource.

WETLANDS OF NORTHERN AUSTRALIA

The water cycle

The freshwater that is vital for life on land, economic development, and farming is endlessly recycled. The process begins with water evaporating from seas, lakes, and forests to form clouds (see panel, opposite). When rain falls, the water is stored in forests, soil, and rocks to be released into rivers and lakes. Some is stored as snow, which melts in spring and summer, enabling rivers to flow during otherwise dry periods. Different human impacts, including deforestation, climate change, and soil damage, are interfering with how the water cycle works, with resulting water shortages in some parts of the world including North Africa and the Middle East.

4 Clouds are formed from water droplets or ice crystals depending on the temperature. In cooler temperatures precipitation falls as snow.

5 Water droplets collide and merge in clouds and fall as rain, sleet, snow, or hail.

Glaciers store water that is released as snowpack melts in the summer. The loss of glaciers due to climate change is a water security issue.

6 Water sinks into the soil in a process called infiltration. The process is assisted by intact vegetation and roots

7 Some of the water that filters into the soil is stored deep beneath the surface as groundwater. More than 30 per cent of Earth's freshwater is stored as groundwater.

HOW CLOUDS ARE FORMED

Clouds form as warm air is forced upwards. As water condenses in rising air it releases heat. This warms the air mass, and causes it to rise further. The air cools and the relative humidity increases. Rising air gets saturated and water vapour collects around airborne particles to form a cloud.

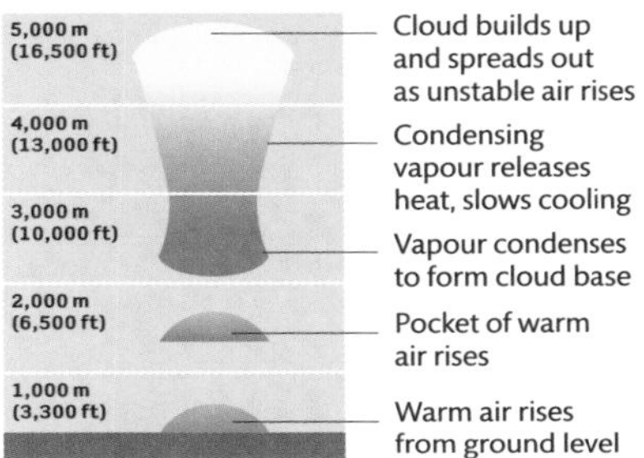

3 As water vapour rises it cools and condenses into water droplets.

2 Plants and trees take in water through their roots. Most of it passes out through the pores in their leaves as water vapour.

Cloud forests harvest water from clouds to create flows of liquid water. The large surface areas of leaves at cool, cloudy altitudes snatch water from clouds and are dripping wet, even when it is not raining.

1 Water is heated by the Sun and turns into vapour. Microscopic plankton release a gas called dimethyl sulphide that hastens the condensation of vapour and "seeds" clouds.

8 Groundwater flows beneath the surface and eventually discharges into the sea, mostly via rivers.

Water footprint

It is not the everyday water that we use at home that forms most of our water consumption. The vast majority we use is "hidden" water needed to grow food, produce goods, and generate energy.

Water resources are more vital to world trade than oil and financial capital. Similar to a carbon footprint (see pp44–45), a "water footprint" shows the extent and location of water used by individuals, businesses, and countries. This allows us to calculate the amount of "virtual" water. This is the water used to make traded goods and helps to reveal which countries rely on freshwater imports to meet their needs – for example, those with limited water resources of their own.

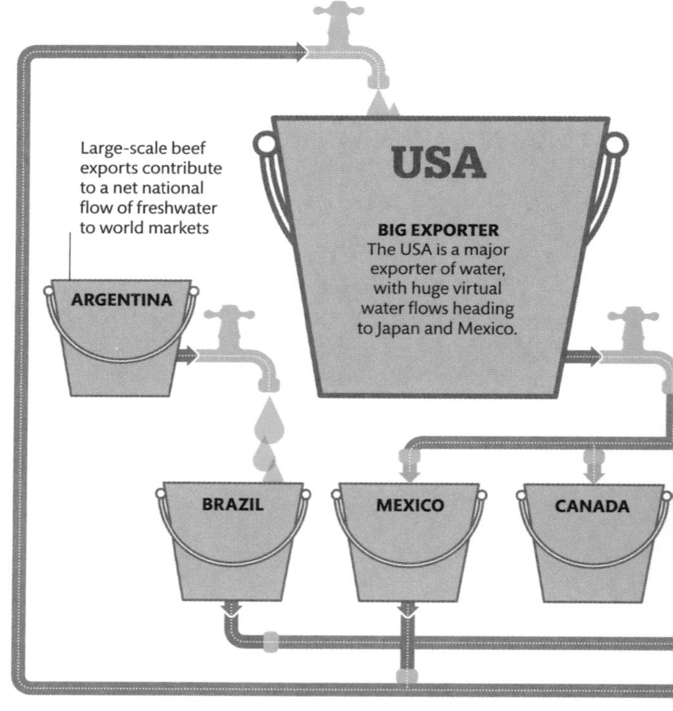
Large-scale beef exports contribute to a net national flow of freshwater to world markets
USA
BIG EXPORTER
The USA is a major exporter of water, with huge virtual water flows heading to Japan and Mexico.
ARGENTINA
BRAZIL
MEXICO
CANADA

Trading virtual water

All countries import and export food, so they all trade virtual water. The volume of water needed to trade agricultural and industrial products from 1996–2005 averaged 2.3 trillion cubic metres (82 trillion cubic feet) per year, or about five times the volume of Lake Erie, North America. Among the biggest net exporters of virtual water are the USA, China, Canada, Brazil, and Australia. Among the biggest net importers are Europe, Japan, Mexico, South Korea, and the Middle East.

HOW MUCH WATER?

Each person in the UK uses an average of 145 litres (32 gallons) of water each day for cooking, cleaning, and washing. When including virtual water, however, this figure rises to a colossal 3,400 litres (748 gallons) per day. Cotton and leather goods have a significant water footprint. The longer these products can be made to last, the lower their overall impact will be.

KEY
100 litres (22 gallons)
1,000 litres (220 gallons)

MICROCHIP
32 litres
(7 gallons)

APPLE
70 litres
(15 gallons)

HAMBURGER
2,400 litres
(530 gallons)

COTTON T-SHIRT
4,100 litres
(900 gallons)

PAIR OF LEATHER SHOES
8,000 litres
(1,760 gallons)

Largest water footprints

The top water-consuming countries include those with both high and low per capita incomes, revealing that freshwater is vital at all stages of economic development. Countries with low rainfall face bigger challenges than wetter ones. Some countries, such as Brazil, rely on rainwater to meet their needs for food production, whereas India uses a lot more river water to irrigate crops for its huge agricultural sector. Roughly two-thirds of China's water footprint is used for agriculture, with one quarter supplying its massive manufacturing sector.

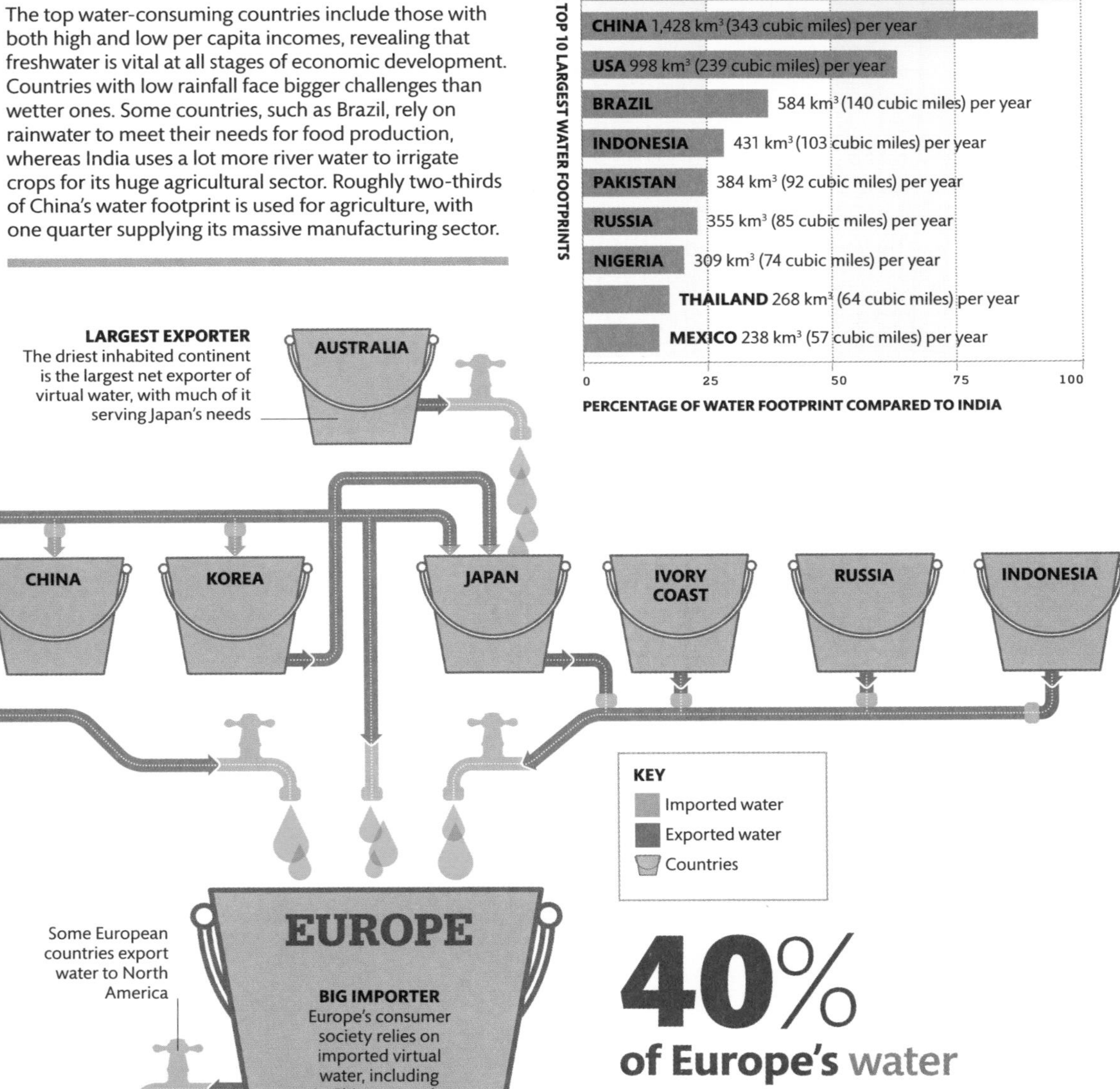

40%

of Europe's water footprint lies outside its borders

Consuming passions

The last century saw a dramatic increase in demand for all kinds of natural resources. In 2020, for the first time, combined consumption of construction materials, ores and minerals, fossil fuels, and biomass reached 100 billion tonnes (110 billion tons), more than 10 times more than in 1900. While rising demand fuels economic growth, it places increasing pressure on natural systems, leading to environmental problems. Unless we adopt different consumption and production patterns, projected population growth and economic development will further intensify pressure on the environment.

Rocketing resources

Every item we use and dispose of originates from natural resources. Some, such as wood used to make paper, are renewable; others, such as minerals, are not. Turning raw materials into products uses energy and water, and creates wastes of different kinds, including carbon dioxide. The world's rocketing demand for resources is rarely seen in the context of the atmospheric and ecosystem damage it causes. Even if such impacts are understood, resource supply is usually prioritized as vital for economic growth.

"There are **constant assaults** on the natural environment, the result of **unbridled consumerism,** and this will have **serious consequences** for the world economy."

POPE FRANCIS

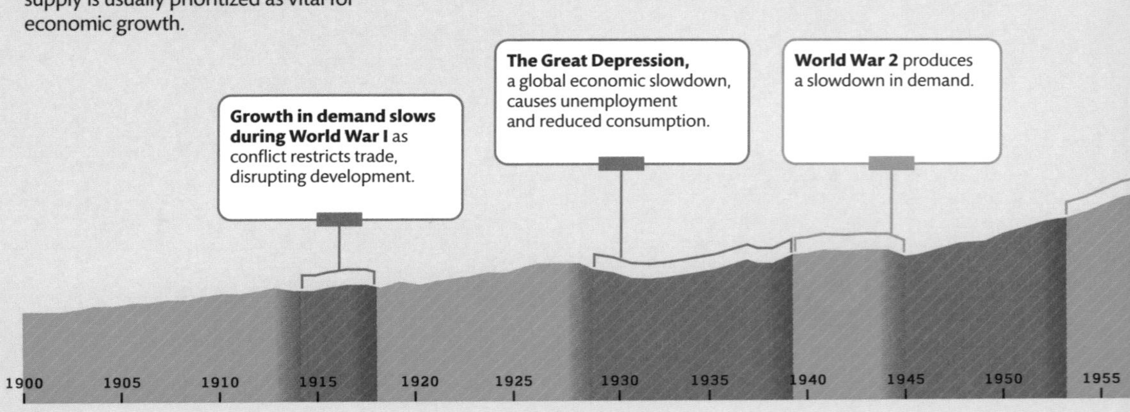

Growth in demand slows during World War I as conflict restricts trade, disrupting development.
The Great Depression, a global economic slowdown, causes unemployment and reduced consumption.
World War 2 produces a slowdown in demand.
1900
1905
1910
1915
1920
1925
1930
1935
1940
1945
1950
1955
YEAR

THE BIG USERS

When it comes to the consumption of natural resources, we are not all equal. Higher-income countries, such as France, Germany, Japan, the UK, and the USA, use more natural resources due to their high-consumption patterns than countries with a lower average income per person. As efforts to stimulate economic growth in poorer countries achieve some level of success, however, consumption patterns there will begin to follow those of wealthier nations.

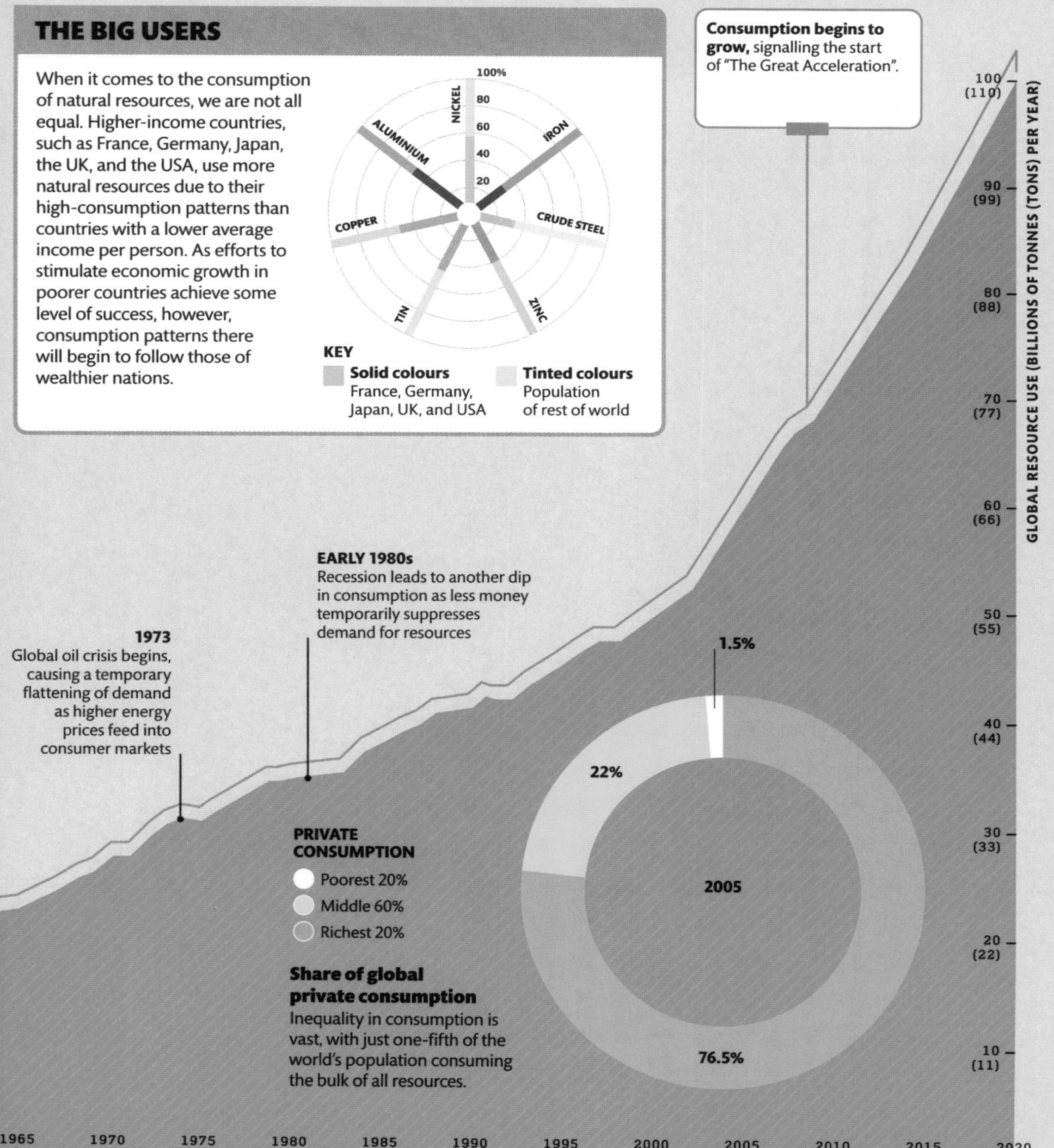

Share of global private consumption
Inequality in consumption is vast, with just one-fifth of the world's population consuming the bulk of all resources.

The rise of consumerism

Rising living standards have led to an explosion in demand for all kinds of consumer goods, ranging from disposable packaging to complex durable products such as cars. All require natural resources – and all eventually become waste.

The spread of middle-class lifestyles has produced a rocketing demand for resources. Bottled water and cars are just two examples that reflect wider trends. Whereas both were once absent from our lives, today they are pervasive, especially in richer countries and those with fast-growing economies.

Rising demand for these and other products puts pressure on limited natural resources such as oil and minerals. Increasing amounts of water and energy are needed for their manufacture, while increased product consumption is adding to global waste. Cleaner, more efficient production methods and the more effective elimination of waste, which can in turn be used to make new products, can diminish the impact of more affluent lifestyles.

Bottling water: the true costs

Bottled water is usually sold in plastic or glass bottles. Extracting the water itself can deplete resources and cause local environmental impacts, but the energy and resources used for its transport and packaging lead to the biggest global effects. Waste caused by plastic bottles is another serious problem.

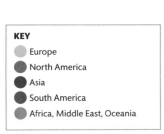

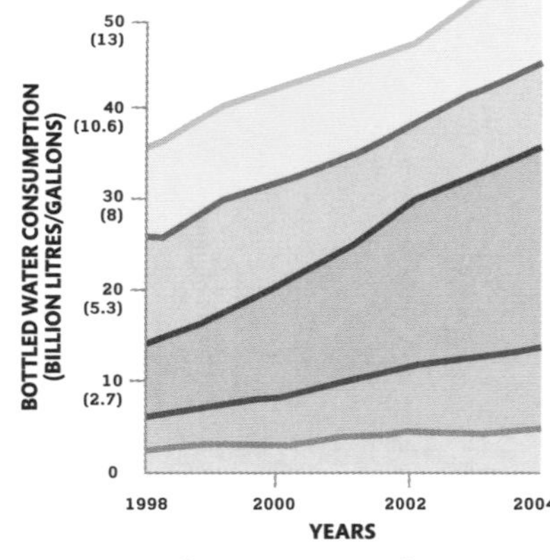

KEY

- Europe
- North America
- Asia
- South America
- Africa, Middle East, Oceania

Increasing consumption

Bottled water sales have vastly increased since the 1990s. By 2017, the total consumed had approached a staggering 400 billion litres (106 billion gallons).

<1% Treatment at plant

<1% Filling, labelling, and sealing bottle

4% Refrigeration

Energy in a bottle

Treating water and filling a bottle with it takes only a tiny amount of energy. Making and shipping the plastic container demand 95 per cent of the total required energy costs.

45% TRANSPORTATION

50% PRODUCTION OF PLASTIC BOTTLE

877 the number of plastic bottles **thrown away every** second

Materials in a car

The process of car manufacturing requires everything from metal ore extraction to applying paint and fitting complex electronics. Making cars also uses huge quantities of energy and water. Manufacturers seek ways to reduce the overall impact of vehicles, not only when driven, but also in production and by recovering materials when cars are scrapped. To that end, some companies are building lighter-weight and more fuel-efficient cars made from recycled aluminium.

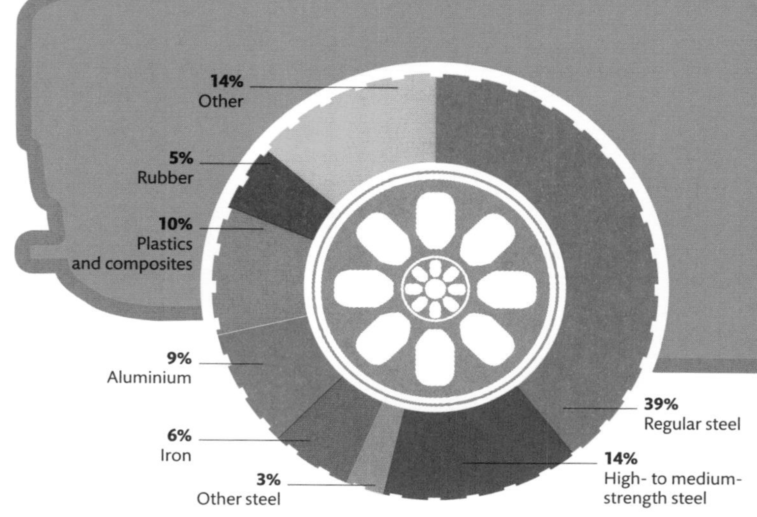

Vehicle ownership

Only in the USA, the world's most mature car market, has the number of cars owned per person recently stabilized, and between 2018 and 2019 the number of registered vehicles in the USA actually fell, by nearly two million, suggesting car ownership there may have peaked.

PASSENGER CAR DENSITY PER 1,000 INHABITANTS

500
400
300
200
100
0

EUROPEAN UNION	JAPAN	USA	SOUTH KOREA	RUSSIA	BRAZIL	CHINA	INDIA
487	463	404	300	259	147	50	13

> "If we want a **sustainable society,** we need to get **consumers** to **think about their purchases."**
>
> **DAVID SUZUKI, CANADIAN SCIENTIST**

Passenger car ownership
The number of passenger cars owned changes rapidly as national economies develop. In 2005, China had just 11 cars per 1,000 people; in 2012, however, that figure had more than quadrupled.

Wasteful world

All the waste we generate originates from natural resources, which are often extracted in environmentally damaging ways. Waste disposal also causes problems, such as pollution and climate change.

World population increase and economic growth have led to an explosion in demand for resources. As the overall level of consumption has risen, there has been a dramatic increase in the amount of waste generated. That waste includes food, wood, metals, construction materials, and plastics, as well as complex high-technology products such as cars and computers. The production of all of these items results in greenhouse gas emissions, and even more are added during the process of disposing of them: for example, rotting food waste in landfill sites releases methane, a very powerful climate-changing gas.

There are three basic approaches to waste management: burying it in the ground, burning it (sometimes with energy recovery technology), or recycling. From an environmental viewpoint, however, the best option is to avoid producing waste in the first place.

Mounting waste

In 1900, the world produced about half a million tonnes (0.55 million tons) of solid waste per day. By 2000, that quantity was six times higher, and by 2100, based on projected population, social, and economic trends, it is expected to quadruple again to about 12 million tonnes. Adopting more ecologically sound consumption patterns and increasing recycling could, however, lead to a much lower daily peak of approximately 9.5 million tonnes by åthe middle of the 21st century.

1900

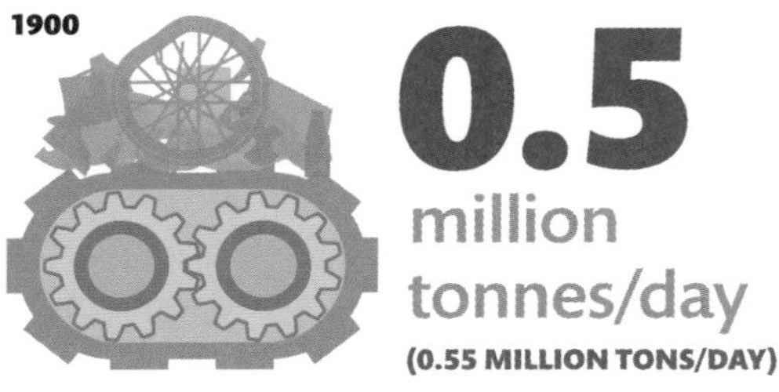

0.5 million tonnes/day

(0.55 MILLION TONS/DAY)

2000

3 million tonnes/day

(3.3 MILLION TONS/DAY)

2100

What's in the bin?

There are huge differences in waste produced in the affluent west and that generated by developing nations. For example, a far higher proportion organic waste is put into bins in Lagos, Nigeria, compared with New York State, USA. New Yorkers waste more plastic, and overall, the American consumers are producing about three times as much waste per person per day as people in Lagos, who generally live on lower incomes.

700 The number of years it can take **a plastic bottle** to break down

TECHNO TRASH

In 2016, about 50 million tonnes (55 million tons) of electronic waste was generated. Computers, mobile phones, and televisions are among the products that comprise this growing mountain.

Japan **16.9kg (37 lb)**
USA **19.4kg (43 lb)**
Russia **9.7kg (21 lb)**
Brazil **7.4kg (6 lb)**
Germany **22.8kg (50 lb)**
UK **24.9kg (55 lb)**
China **5.2kg (11 lb)**
India **1.5kg (3 lb)**
Norway **28.5kg (63 lb)**
Australia **23.6kg (52 lb)**

ELECTRONIC WASTE PER PERSON IN SELECTED COUNTRIES

12 million tonnes/day
(13.3 MILLION TONS/DAY)

Where does it all go?

As our consumption levels rise and we generate increasing quantities of rubbish, the management of solid waste has become an unprecedented – and increasingly important – challenge.

Currently, four main options exist when it comes to the disposal of solid waste material: burying waste in landfill; burning it in different kinds of incinerators, some of which are also capable of generating heat and/or power; recycling; and for organic matter, composting or anaerobic digestion to produce biogas for energy, while also recovering nutrients that would otherwise be lost.

The first two disposal methods are the least environmentally sustainable. The huge diversity of man-made materials, including many types of plastic, that cannot be easily separated and therefore recycled exacerbate the problem. Unfortunately, however, these two options are still viewed as the cheapest and/or easiest solutions for the growing waste mountains being generated by many societies today.

Where waste ends up

The figures presented here are based on data collected on member countries belonging to the Organisation for Economic Co-operation and Development. Each wheel shows the percentage of a particular waste-disposal method used by each country in 2019. This data reveals a very mixed picture with some countries recycling at high levels and others still reliant on landfill.

Landfill
Burying waste in the ground can cause groundwater pollution as toxic substances are released. Rotting organic waste also emits methane, one of the primary greenhouse gases.

Incineration
Burning any kind of waste can cause air pollution. However, burning plastics and other man-made substances also produces residual toxic ash that is frequently buried in landfill.

What can we do?

- **Governments** can set targets to shift more waste to composting and recycling.
- **Governments can** provide incentives to waste operators for change: for example, by taxing landfill waste.
- **Companies can** make packaging and electronic goods more recyclable.

What can I do?

- **Know your waste**. Learn what can be recycled and put it in the correct bin, whether in your home or at a collection facility.
- **Buy with care**. Avoid unnecessary packaging and single-use or disposable items.
- **Avoid plastic bags.** Buy durable shopping bags to use when you go shopping.

Poisoning the Earth
As waste breaks down in landfill, water filters through it, forming a toxic liquid called leachate that can seep into soil and groundwater.

90% The **energy saving** when making an aluminium can from **recycled waste** compared with **ore**

Recycling
Glass, metals, paper, card, and some types of plastic can be recycled into new products. This process takes much less energy than manufacturing the same items from raw materials – and it also saves resources.

Composting
Organic matter such as food waste, agricultural waste, and plant material can be used to make biogas, generating heat and electricity while at the same time saving nutrients that can be returned to the soil as fertilizer.

Chemical cocktail

The number of man-made chemical substances being released into the environment is increasing dramatically. We don't yet know the impact they may have, including any "cocktail effects" if two or more combine.

Persistent organic pollutants (POPs) are mostly man-made compounds that do not readily degrade or break down in the environment. Because of this, they last a long time and accumulate in food chains, causing serious biological effects, especially among larger organisms. POPs include many chemicals that were developed as beneficial substances, such as the insecticide DDT and the PCBs once used in electrical equipment. Others, such as dioxins, are created via combustion, such as by burning waste in incinerators.

THE RISE IN NEW CHEMICALS

Since the 1940s, millions more synthetic compounds have been invented, registered, manufactured, and released into the environment. Many have not been properly assessed for their biological impacts, either on their own or in combination with other substances.

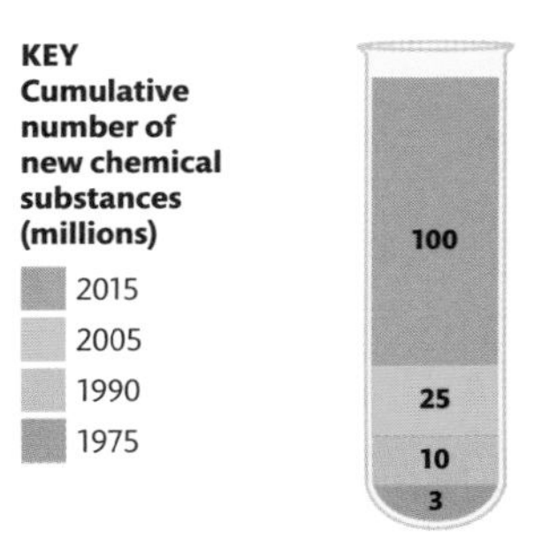

What is biomagnification?

As POPs move through food webs, they become more concentrated as one species feeds on another. For example, after the (now banned) insecticide DDT entered lakes and other water sources, it built up in the bodies of top predators such as fish-eating ospreys, causing them to lay thin-shelled eggs that broke when adult birds incubated them.

DDT runs off fields in rainwater
Once applied, DDT enters bodies of water, such as rivers, lakes, and reservoirs, at about 0.000003 parts per million (PPM).

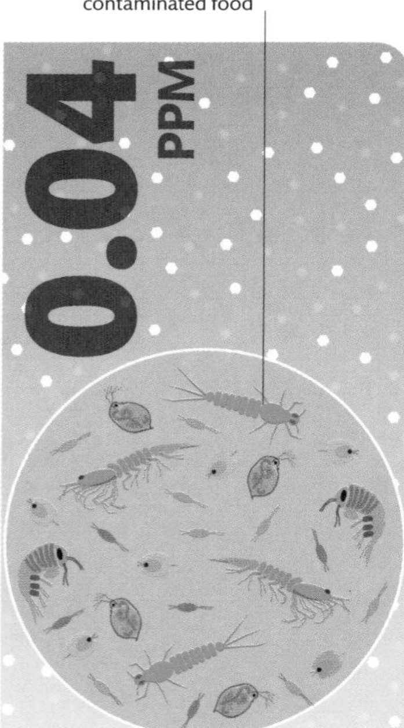

Small creatures consume DDT
Zooplankton, tiny creatures that live in water, consume microscopic food items contaminated with DDT and their bodies accumulate the chemical to around 0.04 PPM because the substance does not break down once eaten.

What can we do?

- **Governments can work together** to control the effects of chemicals, such as via the Stockholm Convention on Persistent Organic Pollutants.
- **Governments can** introduce more rigorous testing regimes to reveal the potential biological effects of new and existing chemicals.

What can I do?

- **Reduce your exposure** to potentially harmful substances. Start by looking up what is listed on the labels of consumer goods.
- **Join campaigns** that support regulating chemicals entering the environment and advocate more effective screening of new substances.

Small fish feed on the plankton
As small fish eat small creatures contaminated with DDT, they further concentrate the DDT to around 0.5 PPM. The DDT is lodged in the fishes' bodies but does not break down; it continues to accumulate in larger amounts.

Predatory fish
Larger fish, such as trout, that eat smaller fish have higher concentrations of DDT in their bodies at around 2 PPM. These fish become food for top predators, such as bears, fish-eating birds, and ultimately humans.

DDT reaches top of the food chain
At around 25 PPM – about 10 million times more concentrated than when the chemical first entered the water – this amount threatens the survival of many species; for example, bald eagle populations were wiped out in much of North America when DDT was used.

"As **human beings we are more urbanised than ever before,** and we are out of touch with the natural world. Yet we are **100% dependent on its resources."**

SIR DAVID ATTENBOROUGH, BRITISH BROADCASTER AND NATURALIST

The global age

Better lives for many

Our changing atmosphere

Changing the land

Sea changes

The great decline

2 CONSEQUENCES OF CHANGE

Some aspects of rapid change are positive, but others are causing negative consequences for people and the natural world, including the impacts of climate change, pollution, and land degradation.

The global age

Our world is more interconnected than ever before. People can share information, ideas, and images between computer devices all over the world. Airliners fly millions of travellers to cities huge distances apart every day. Once the domain of a small elite, access to cheap travel, high-speed internet, and mobile communications is now growing fastest in developing countries. Interconnection speeds up economic growth, and shapes all forms of business.

Rise of the internet

In 1989, English inventor Tim Berners-Lee devised the world-wide web, which kick-started an information revolution. Events can be watched anywhere in real time, while email offers cheap communications for anyone with an internet connection. Home internet connections became available in the 1990s, and each year many millions more people joined the global digital community. By 2005, there were one billion internet users. This doubled in just five years, and by 2021 had doubled again to reach about four billion. This graph charts the incredible rate of expansion, with about half the global population having access to an internet connection via a home computer or mobile device.

"We must **make globalization an engine** that **lifts people out of hardship and misery,** not a force that holds them down."

KOFI ANNAN, FORMER UNITED NATIONS SECRETARY-GENERAL

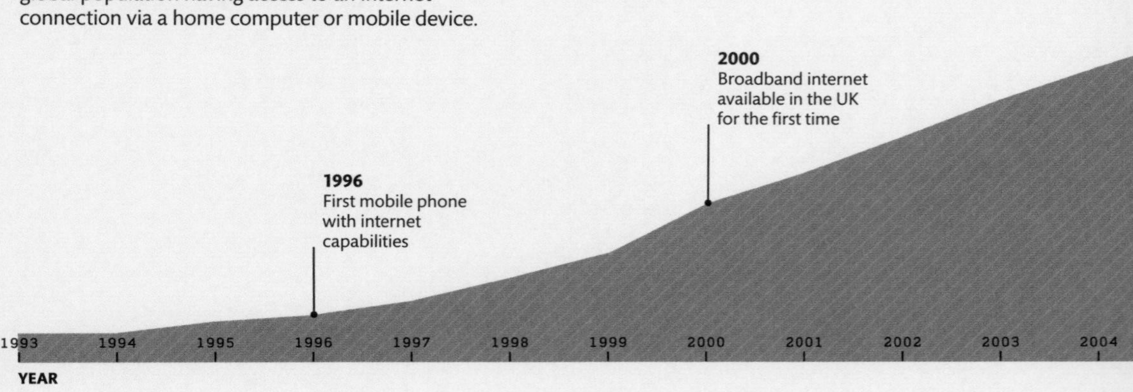

ENHANCING ECONOMIES

Internet access has had a positive impact on economies across the world. The ability to deliver company information quickly, widely, and cheaply means businesses can share corporate news, provide flexible working, lead in innovation, and effectively manage finance. The internet has also diminished the power of established media, allowed social movements to spread their messages and empowered research communities to share data.

INTERNET BENEFITS
INTERNET ACCESS
Increased access
INFORMATION FLOWS
Enhanced speed
INNOVATION
New business ideas
FINANCIAL CAPITAL ACCESS
Enhanced access
ENTREPRENEURSHIP
Facilitation of entrepreneurship
LABOUR ENHANCEMENT
Improved human resources

N America
88
S Asia
30
Sub-Saharan Africa
25
Latin America and Caribbean
66
East Asia and Pacific
56
Europe and Central Asia
80
0%
100%

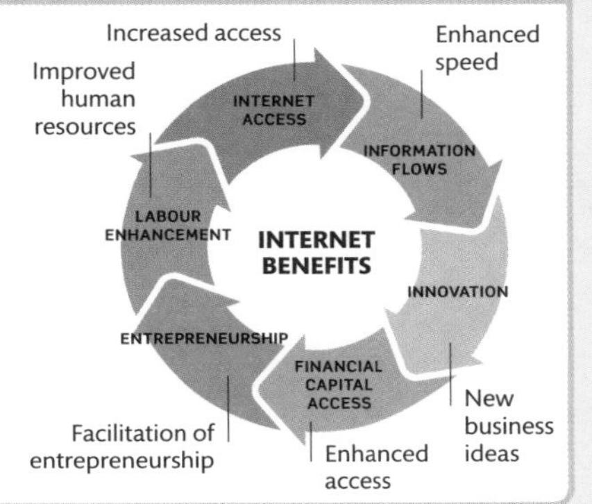

Global internet usage
With rapid growth in both population and economic prosperity, in 2019, in most regions over half the population had internet access.

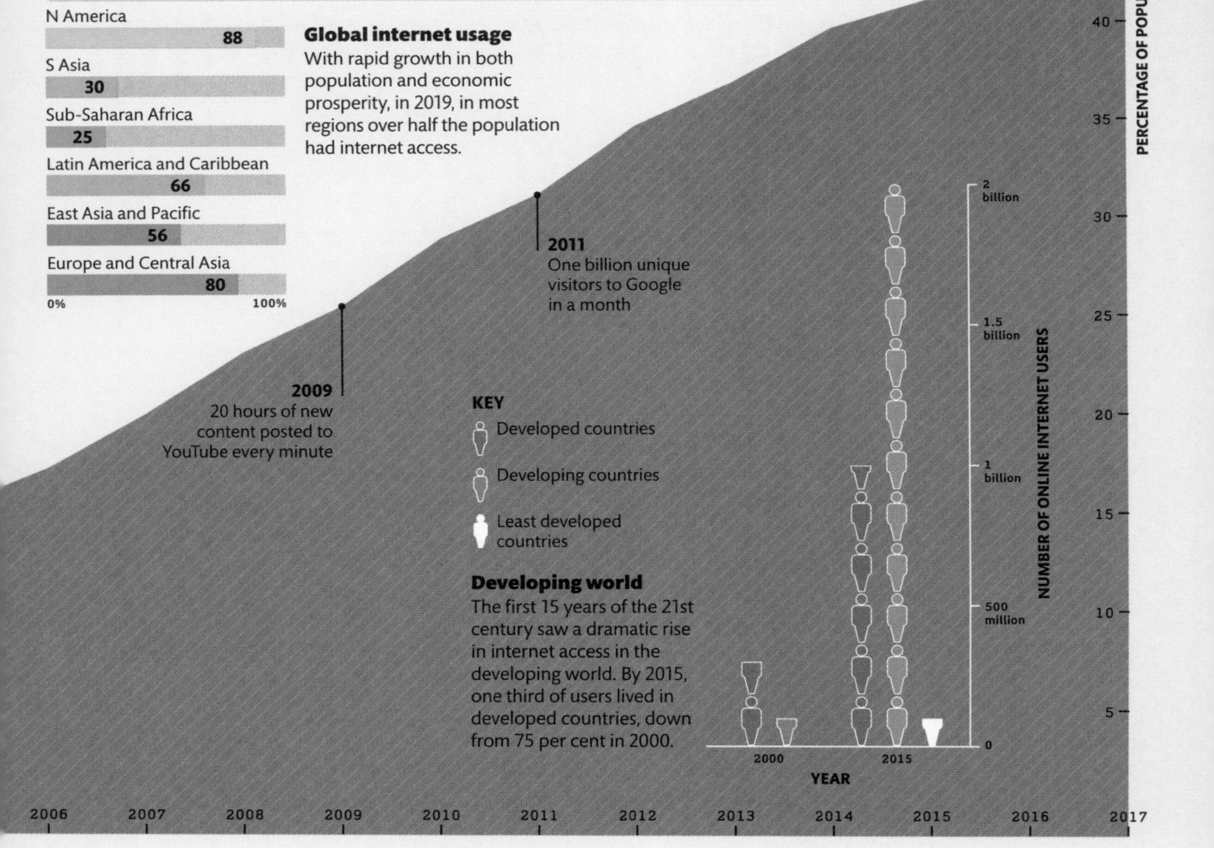

2009
20 hours of new content posted to YouTube every minute
2011
One billion unique visitors to Google in a month
KEY
Developed countries
Developing countries
Least developed countries
2 billion
1.5 billion
1 billion
500 million
0
NUMBER OF ONLINE INTERNET USERS
2000
2015
YEAR
50
45
40
35
30
25
20
15
10
5
PERCENTAGE OF POPULATION WITH INTERNET
2006
2007
2008
2009
2010
2011
2012
2013
2014
2015
2016
2017

Developing world
The first 15 years of the 21st century saw a dramatic rise in internet access in the developing world. By 2015, one third of users lived in developed countries, down from 75 per cent in 2000.

Mobile connectivity

Today, mobile phones are ubiquitous all over the world – from the biggest cities to remote villages – as more and more people connect to the virtual grid to make calls, send texts, and use the internet.

Mobile phones have changed from being a bulky luxury to an everyday item. The first mobile phone was developed in 1973, but it didn't become commercially available for another 10 years, when it was sold for US$4,000 – equivalent to almost US$10,000 in 2015 prices. Many thought it a pricey gimmick.

At the turn of the century, mobile use was concentrated in Europe and North America, but usage has rocketed all over the world as the technology has become cheaper. The instant communication and information that mobile technology brings is transforming the way people live their lives. Mobiles are no longer just a means of voice communication but give users access to banking, healthcare, and the political process.

Remote connectivity
Nomadic peoples such as this Maasai warrior on the Kenyan plains now have ready access to mobile communication.

Upwardly mobile

In the 20 years to 2013, all global regions saw a massive expansion in mobile phone use. The fastest growth was in Latin America and the Middle East, where phone masts were constructed instead of the cables and wires needed for landline connections, thereby "leapfrogging" old technology. In several regions, the number of mobile connections increased to more than 100 per cent of the total available market.

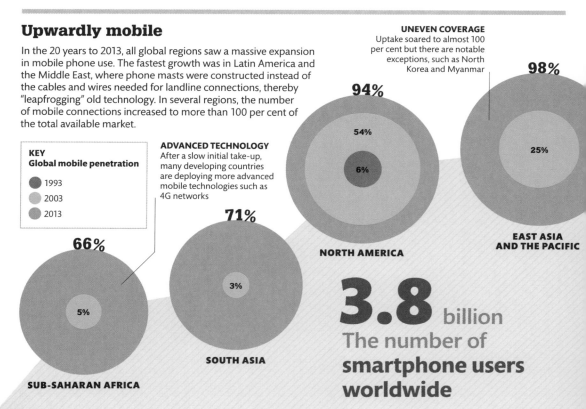

KEY
Global mobile penetration

- 1993
- 2003
- 2013

ADVANCED TECHNOLOGY
After a slow initial take-up, many developing countries are deploying more advanced mobile technologies such as 4G networks

UNEVEN COVERAGE
Uptake soared to almost 100 per cent but there are notable exceptions, such as North Korea and Myanmar

SUB-SAHARAN AFRICA: 66% (2013), 5% (2003)

SOUTH ASIA: 71% (2013), 3% (2003)

NORTH AMERICA: 94% (2013), 54% (2003), 6% (1993)

EAST ASIA AND THE PACIFIC: 98% (2013), 25% (2003)

3.8 billion
The number of **smartphone users worldwide**

Device explosion

The first mobile phones were out of reach for all but the very wealthy, but as demand grew, prices fell and uptake exploded. At the same time as devices became cheaper, their features increased, leading to the current success of the smartphone. Ongoing improvements in signal coverage, battery life, and handset size have helped drive their popularity. The rise of the smartphone has supported a range of social revolutions, including the rise of online shopping.

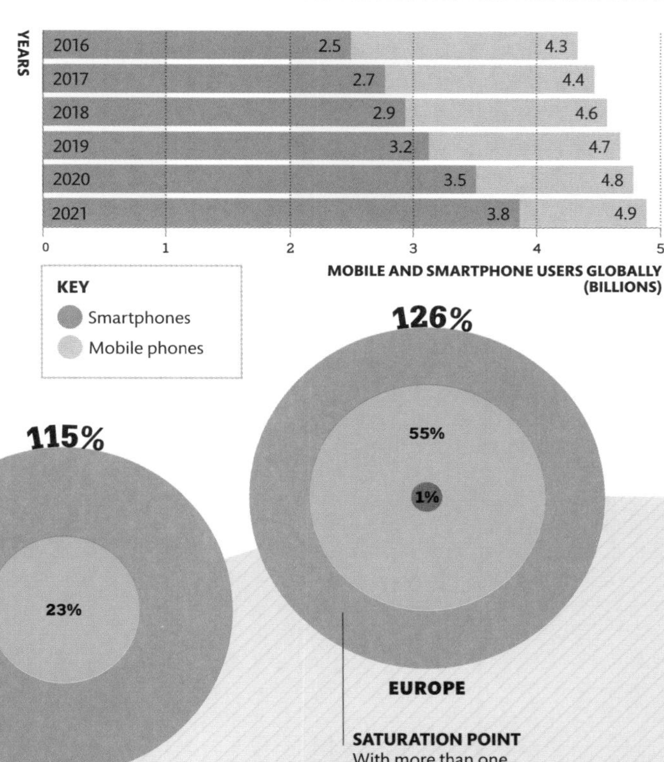

SATURATION POINT
With more than one mobile phone per person in Europe, the growth of the mobile market is likely to slow down here

SOCIAL MEDIA USE AROUND THE WORLD

The enhanced connectivity provided by smartphones has been a hugely important driving force behind the rapid rise of social media, with nearly 99 per cent of the users of Facebook, Twitter, Instagram, and other platforms connecting via handsets. The rise of social media has created fundamental changes in many spheres, including in politics and elections, news reporting, and how we gain information.

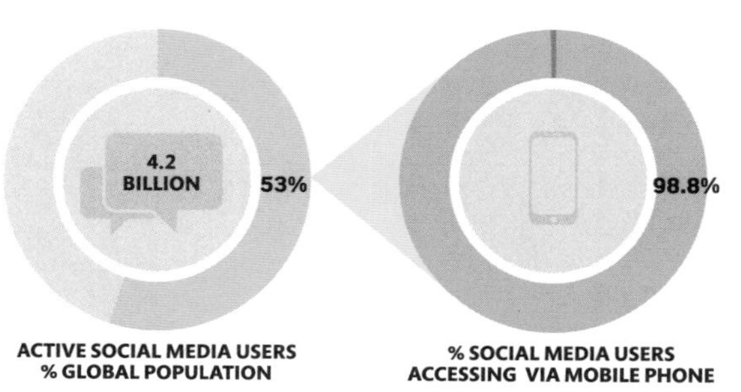

Taking to the skies

The spectacular rise of air travel connects the world as never before. Modern aircraft permit cheap, long-distance transport, making air travel accessible to millions of people and driving economic growth.

The first purpose-built passenger aircraft took to the skies in the 1920s, and the first commercial jet airliners were introduced in the 1950s. Since then, passenger numbers have grown almost yearly as more routes have opened and become more affordable, and aircraft technology has continued to improve. Today, modern aircraft can carry several hundred people. In 2014, there were more than 30 million commercial flights, with the result that about half a million people were in the air at any one time. A network of major airports now connects the globe. The world's busiest airport is Hartsfield-Jackson airport in Atlanta, Georgia, which handled more than 110 million passengers in 2019.

Growth of air travel

In 1970, some 300 million passenger journeys were made by plane. By 2019, this figure had increased fifteenfold to 4.5 billion. This explosive growth was largely a result of rapidly falling costs, which enabled more people to take foreign holidays and also permitted changes to business practices, with more face-to-face contact over long distances. The key drivers of reduced costs in the aviation sector were the removal of monopolies on some routes and more reliable and efficient technology.

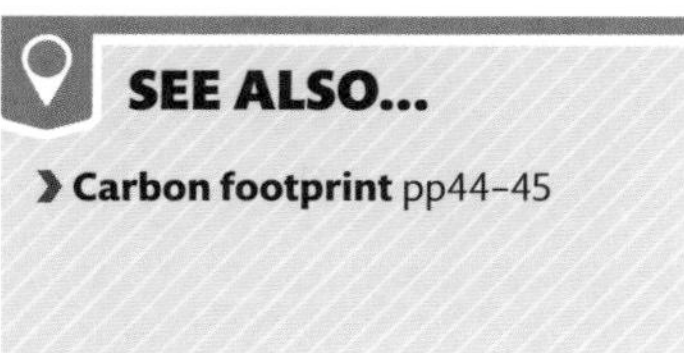

SEE ALSO...

› **Carbon footprint** pp44–45

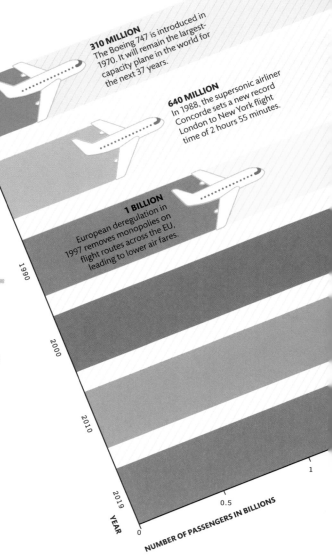

310 MILLION
The Boeing 747 is introduced in 1970. It will remain the largest-capacity plane in the world for the next 37 years.

640 MILLION
In 1988, the supersonic airliner Concorde sets a new record London to New York flight time of 2 hours 55 minutes.

1 BILLION
European deregulation in 1997 removes monopolies on flight routes across the EU, leading to lower air fares.

1970
1980
1990
2000
2010
2019
YEAR

0
0.5
1
NUMBER OF PASSENGERS IN BILLIONS

TOP FLIGHT ROUTES

The most popular flight routes in 2016 were all domestic, and four out of the top five were within the borders of South Korea, Japan, and China. This is because the rapid growth of a relatively affluent middle class in Asia has led to increased demand for flights, including short-haul pleasure trips. The most popular route was between the South Korean capital, Seoul, and Jeju, a holiday island in the south of the country.

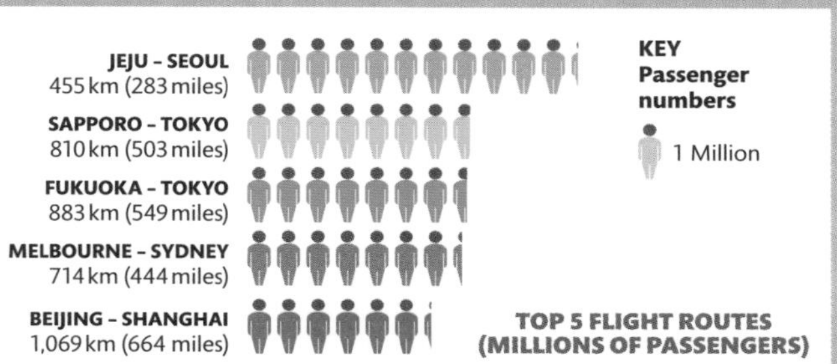

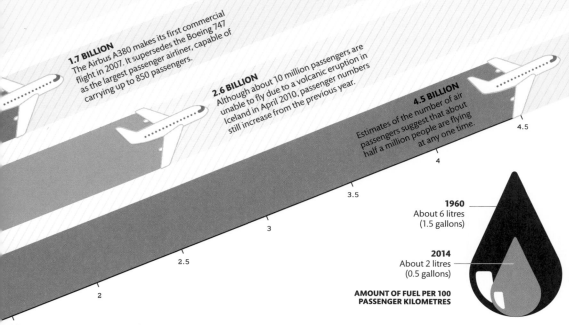

1.7 BILLION
The Airbus A380 makes its first commercial flight in 2007. It supersedes the Boeing 747 as the largest passenger airliner, capable of carrying up to 850 passengers.

2.6 BILLION
Although about 10 million passengers are unable to fly due to a volcanic eruption in Iceland in April 2010, passenger numbers still increase from the previous year.

4.5 BILLION
Estimates of the number of air passengers suggest that about half a million people are flying at any one time.

1.5 2 2.5 3 3.5 4 4.5

1960
About 6 litres (1.5 gallons)

2014
About 2 litres (0.5 gallons)

AMOUNT OF FUEL PER 100 PASSENGER KILOMETRES

Air transport has reduced its fuel use and CO_2 emissions per passenger kilometre **by well over 70% compared to the 1960s**

Increased fuel efficiency

Rising fuel costs and pressure over environmental concerns, especially about air pollution, noise, and climate-changing emissions, have stimulated manufacturers to develop more efficient aircraft. As a result, the fuel needed to fly one passenger 100 km (60 miles) has decreased by more than two-thirds since the 1960s, with a similar reduction in climate-changing emissions.

Better lives for many

There has been significant progress in reducing extreme poverty over recent decades, partly due to economic growth. Access to education, connection to electricity, and the provision of healthcare, clean water, and sanitation all reduce poverty. By helping people escape poverty and improving the economy, these factors create a beneficial cycle for the whole of society. However, while there is global improvement, some parts of the world remain affected by war, conflict, and inequality, which means there is still plenty of work to be done in ensuring better lives for all.

PERCENTAGE OF THE WORLD POPULATION LIVING ON LESS THAN US$1.90 A DAY

40
30
20
10

YEAR

1981 1984 1987 1990 1993 1996

PER CENT OF THE POPULATION

100 90 80 70 60 50 40 30 20 10 0

1981 2010

Low-income countries
Although fewer people are living in extreme poverty at a global level, the proportion experiencing the effects of very low incomes in the poorest countries is still high at around 45 per cent.

KEY
Population living in extreme poverty

Globally

Low-income countries

1992
The Earth Summit on sustainable development in Rio de Janeiro highlights the need to eradicate poverty

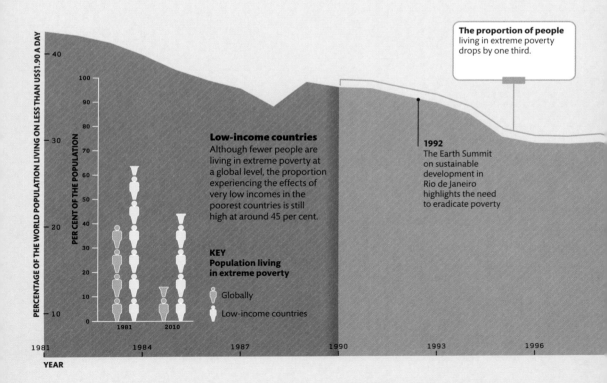

The proportion of people living in extreme poverty drops by one third.

Declining poverty

During the last three decades, the number of people living in extreme poverty has declined significantly. Until 2015, extreme poverty was defined as living on less than US$1.25 a day – the level at which basic survival conditions can be met. In that year, this figure, called the poverty line, was raised to US$1.90.

This reduction in extreme poverty happened despite large population growth during the same period. It occurred because of the steady expansion of countries' economies, leading to increases in average per-capita incomes in both developed and developing nations. The steepest decline began in 1997, when explosive economic growth took off in Asia – particularly in China. This rapid reduction in extreme poverty masked the effect of the two regions where poverty increased – in Eastern Europe and Central Asia after the fall of communism.

WHERE WILL THE POOREST PEOPLE BE IN 2030?

A 2015 ranking that compared countries based on income and the cost of living found the 10 poorest nations in the world were all in Africa. The greatest absolute numbers of people in extreme poverty, however, are mainly located in Asia, as this is where the most populated countries are. Millions of people live in vast slums and large rural populations survive from subsistence farming, all living on tiny incomes.

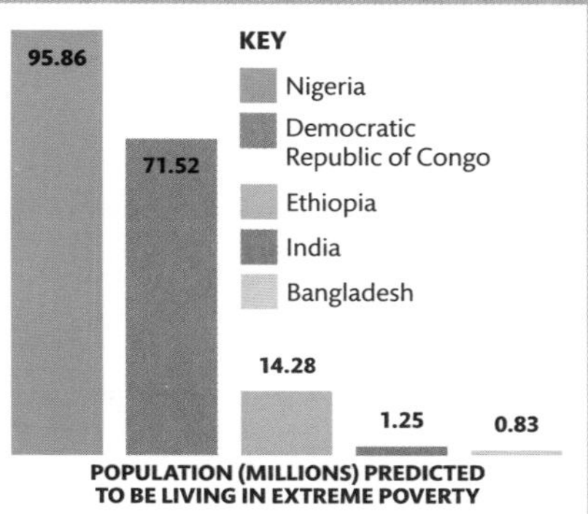

"Saving our planet, lifting people out of poverty, advancing economic growth... these are one and the same fight."

BAN KI-MOON, FORMER UNITED NATIONS SECRETARY-GENERAL

2000
Millennium Development Goals are adopted by the United Nations with, among other things, the aim of reducing hunger and poverty

2005
G8 countries agree to write off the debts of the poorest countries

Sustained economic growth helps nearly half a billion people escape living in extreme poverty.

2002 2005 2008 2011 2014 2017

Clean water and sanitation

Clean water and sewage treatment facilities are key factors in influencing outcomes for public health, development, and poverty. Impressive progress has been made in extending these basic necessities to billions more people.

Improved access to clean water

According to World Health Organization data collected over 22 years, the countries below have made the greatest progress in the world and respective regions by supplying a greater proportion of their citizens with access to safe and clean drinking water. Disparities remain, however, between rural and urban areas, and more people living in the countryside are still unable to make use of reliable water supplies than those residing in towns and cities. Despite recent positive progress, millions still die each year from diseases spread in dirty water. Asia and Africa remain the areas where people are at greatest risk of water-borne diseases.

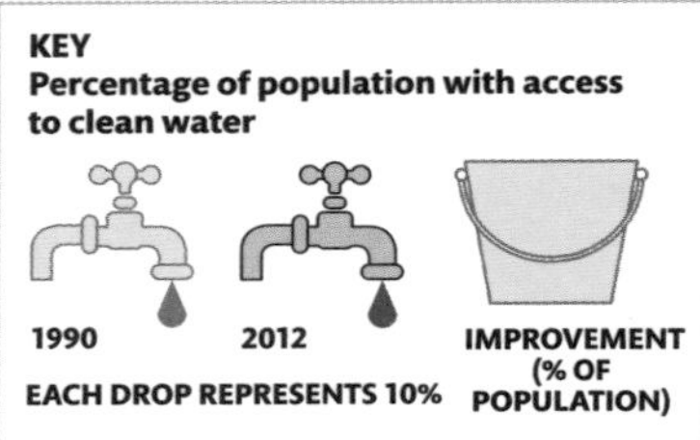

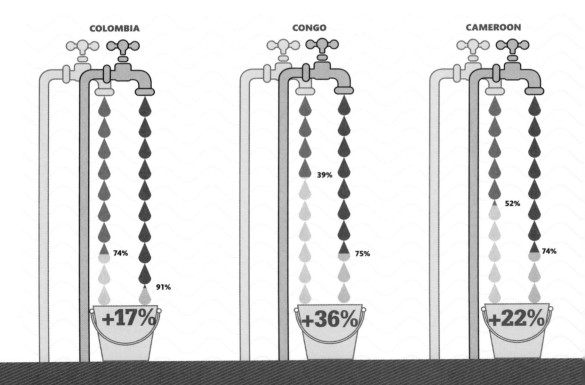

COLOMBIA
74%
91%
+17%

CONGO
39%
75%
+36%

CAMEROON
52%
74%
+22%

Cleaning up water is often the quickest and most cost-effective way to improve public health, saving both lives and money. Following a global improvement programme, around 91 per cent of the world population now has access to safe drinking water access – up by 2.6 billion people compared with 1990. A parallel effort in sanitation means that, today, 68 per cent of the global population has improved sewage treatment and disposal services – up by 2.1 billion compared with 1990. In 2015, however, 2.4 billion people lacked access to basic sanitation facilities. Nearly one billion people are still forced to defecate outside, causing the spread of diseases such as cholera, diarrhoea, and hepatitis A.

1 in 9 people in the world lack access to safe water

Safe to drink
In India, 70 per cent of people had clean water supplies in 2012, leaving 30 per cent still using untreated sources.

IRAQ

69%

85%

+16%

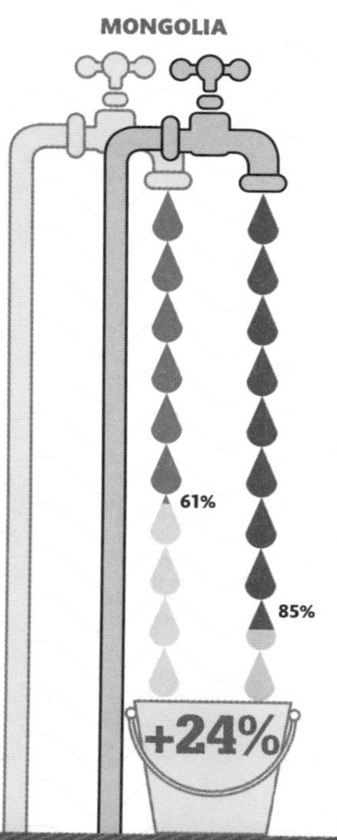

ACCESS TO SANITATION

The stark differences in the improved sewage treatment of the selected countries below reveal their contrasting national circumstances, including the level of development, the rate of economic growth, and the prevalence of corruption.

KEY Percentage of population with access to sanitation

1990

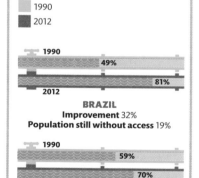

2012

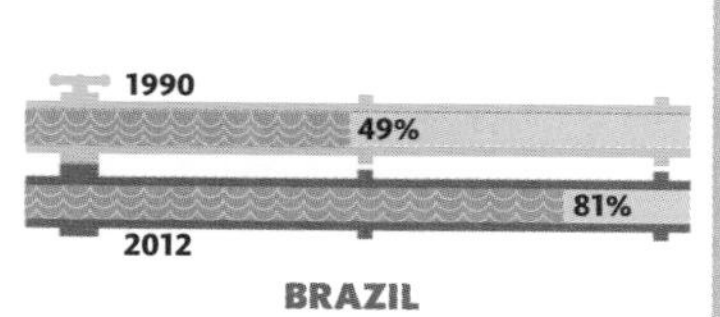

BRAZIL
Improvement 32%
Population still without access 19%

1990
59%
70%
2012

RUSSIA
Improvement 11%
Population still without access 30%

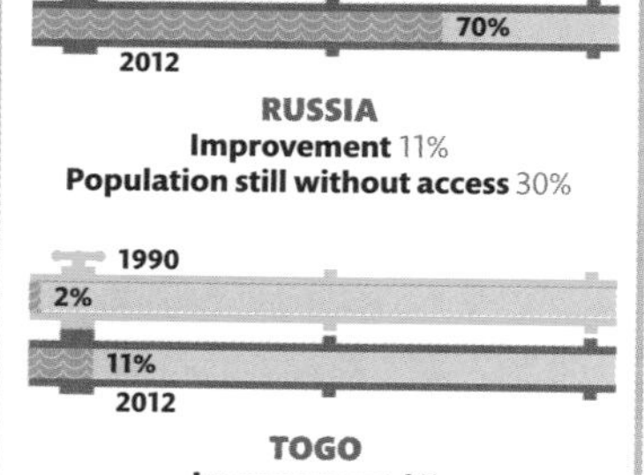

TOGO
Improvement 9%
Population still without access 89%

Reading and writing

Improving literacy skills is essential when trying to reduce poverty. While positive progress has been made in increasing the proportion of people who can read and write, major challenges remain, especially in Africa.

In 2011, there were still 774 million adults in the world who lacked basic literacy skills. Three-quarters of them were living in South Asia, the Middle East, and sub-Saharan Africa, and two-thirds were women.

The past 30 years has seen substantial efforts from governments, charities, and individuals to improve literacy in the poorest and most deprived parts of the world. The ability to read and write greatly improves people's prospects to enter employment, generate income, and contribute to development.

The challenge in achieving universal literacy begins with the acquisition of basic skills during childhood and access to primary education. This was one focal point for the Millennium Development Goals – a set of eight goals set out in a UN initiative in 2000 – and today 91 per cent of children receive primary schooling.

How the world reads

North America, Europe, and Central Asia have all achieved near-universal literacy. The situation in South America has improved during recent decades to reach an average literacy rate of 92 per cent, although the Caribbean still lags behind with just 69 per cent of adults able to read and write. The lowest literacy rates are in sub-Saharan Africa, the Middle East, and South Asia.

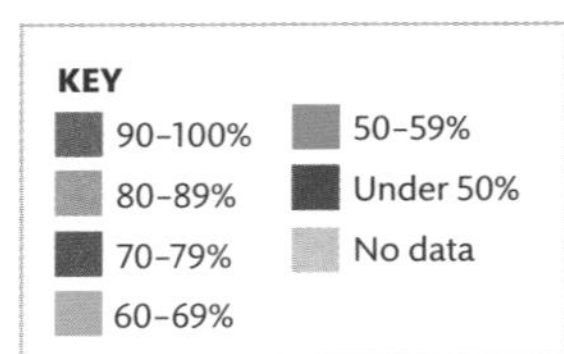

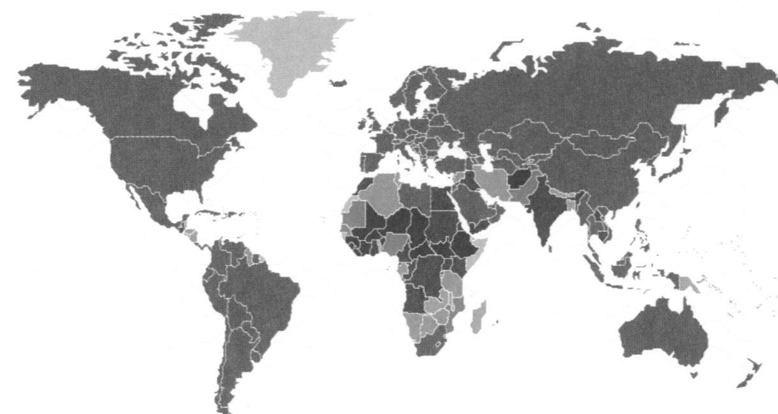

Women's literacy rates

In four of the worst-performing countries, the female literacy rate is less than half that of the male population. In Niger, only one in nine women have basic literacy skills, while literacy in the male population is three times higher. This disparity makes other challenges far harder to address: for example, posing a fundamental barrier to both reducing poverty and slowing population growth (see p22).

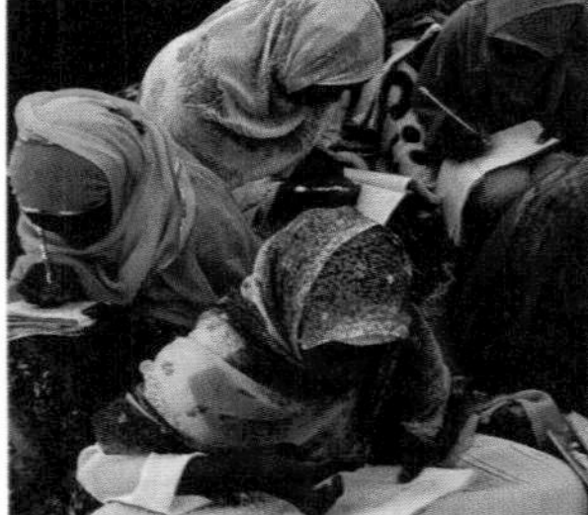

Reading benefits
These women and girls are some of the lucky few who get taught to read and write in Somalia. Here only 25 per cent of women can read and write compared with almost 50 per cent of men.

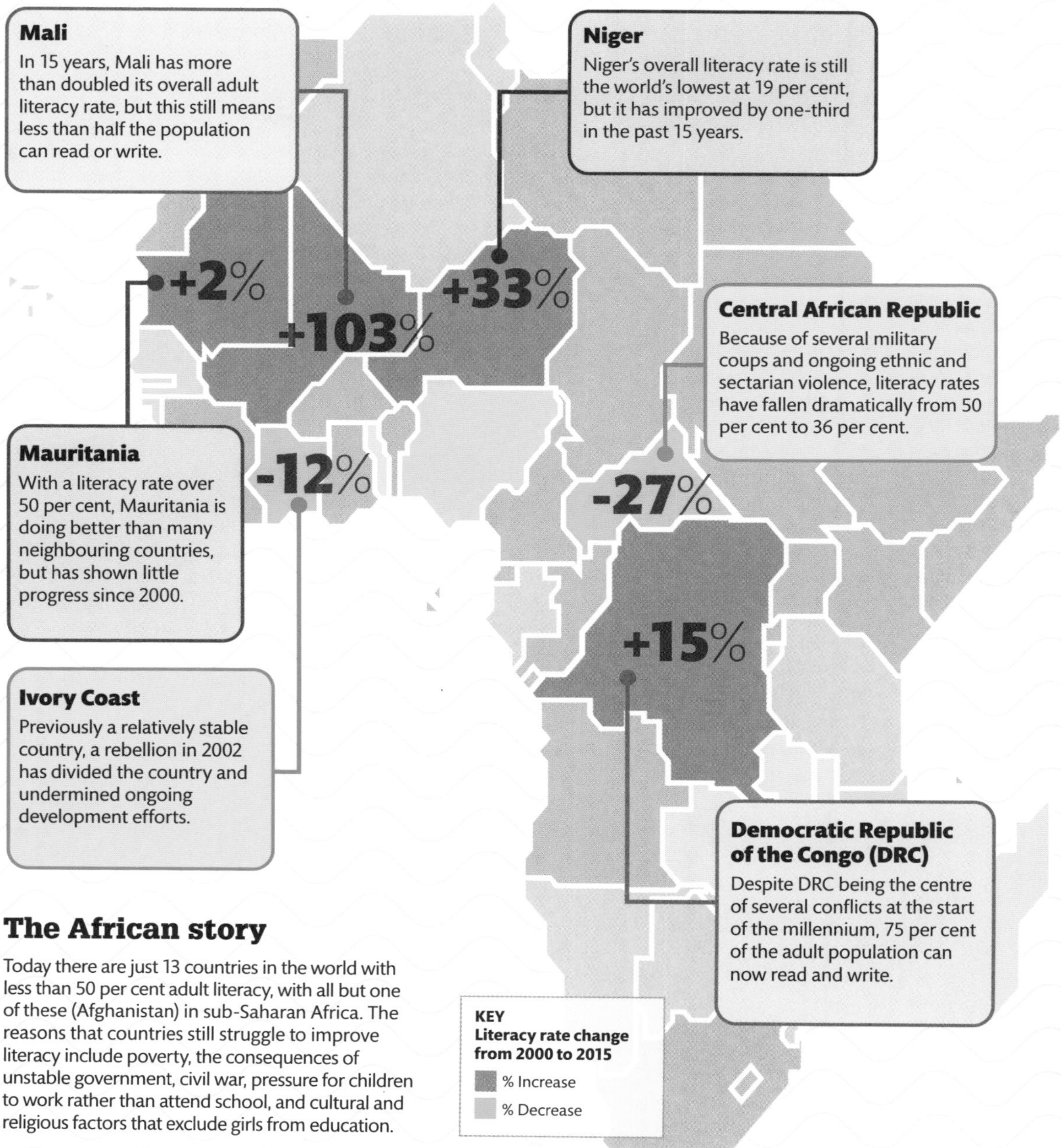

Mali

In 15 years, Mali has more than doubled its overall adult literacy rate, but this still means less than half the population can read or write.

Niger

Niger's overall literacy rate is still the world's lowest at 19 per cent, but it has improved by one-third in the past 15 years.

Central African Republic

Because of several military coups and ongoing ethnic and sectarian violence, literacy rates have fallen dramatically from 50 per cent to 36 per cent.

Mauritania

With a literacy rate over 50 per cent, Mauritania is doing better than many neighbouring countries, but has shown little progress since 2000.

Ivory Coast

Previously a relatively stable country, a rebellion in 2002 has divided the country and undermined ongoing development efforts.

Democratic Republic of the Congo (DRC)

Despite DRC being the centre of several conflicts at the start of the millennium, 75 per cent of the adult population can now read and write.

The African story

Today there are just 13 countries in the world with less than 50 per cent adult literacy, with all but one of these (Afghanistan) in sub-Saharan Africa. The reasons that countries still struggle to improve literacy include poverty, the consequences of unstable government, civil war, pressure for children to work rather than attend school, and cultural and religious factors that exclude girls from education.

KEY
Literacy rate change from 2000 to 2015

- % Increase
- % Decrease

Healthier world

In the 21st century, the incidences of deadly communicable diseases have fallen dramatically, so people lead, on average, longer lives. The major causes of death are now cardiovascular diseases and cancers.

Between 2000 and 2015, the mortality rate in Africa dropped by more than one third, largely due to a reduction in deaths from communicable diseases (those spread from one person to another), including HIV/AIDS. During that same period, deaths caused by malaria in Africa were cut by nearly one half. This was due to simple measures being introduced, such as the increased availability of insecticide-treated mosquito nets and greater access to life-saving medication.

Since 1990, there has been a 44 per cent reduction in maternal deaths around the world, although 830 women still die each day due to complications in pregnancy and childbirth. The success in preventing and treating communicable diseases and avoiding mortality through better public health services has led the causes of sickness and death to shift more towards age- and lifestyle-linked problems, especially cardiovascular and cancer-related conditions.

Major causes of death

The reduction in the death rate across almost all regions means that fewer people are dying each year and, on average, they are living longer. Injuries are responsible for a large number of deaths in Africa, with a proportion far greater than anywhere else in the world. Deaths by non-communicable diseases have stayed relatively consistent across the world.

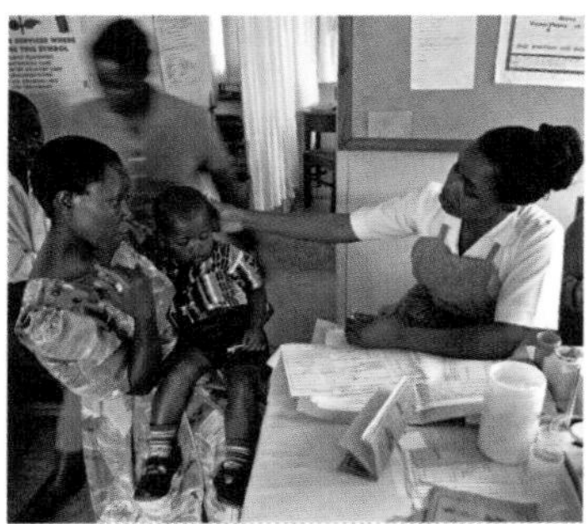

HIV clinic
A nurse comforts a boy diagnosed as HIV positive at a clinic in Kampala, Uganda. Medical investments have reduced mortality from communicable diseases.

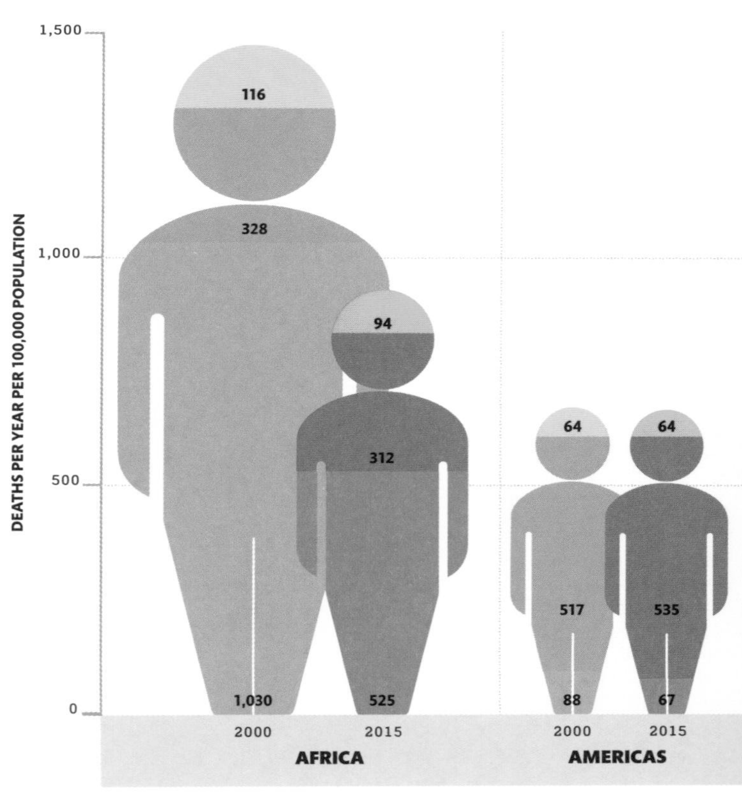

DISEASE AND INCOME

Despite recent improvements in the prevention and treatment of many infectious diseases, the leading causes of death in the world's poorest countries are lower respiratory infections, including pneumonia, bronchitis, and tuberculosis. In the richest countries, one of the fastest-growing causes of death is Alzheimer's disease and dementia, reflecting the increased longevity in the developed world. This puts greater long-term pressure on health services already under strain.

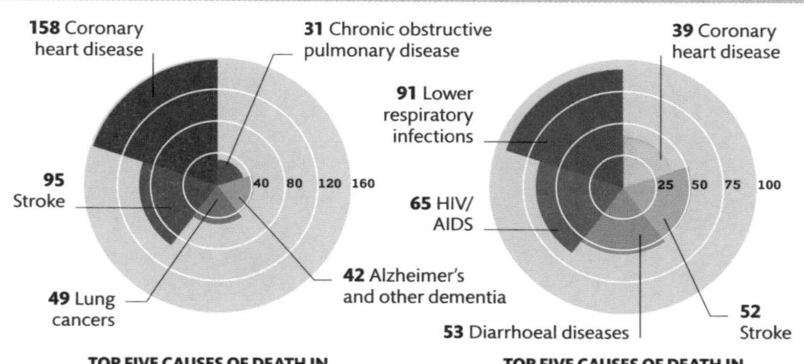

TOP FIVE CAUSES OF DEATH IN HIGH-INCOME COUNTRIES (DEATHS PER YEAR PER 100,000 POPULATION)

TOP FIVE CAUSES OF DEATH IN LOW-INCOME COUNTRIES (DEATHS PER YEAR PER 100,000 POPULATION)

KEY Causes of death

- Injury
- Non-communicable diseases
- Communicable diseases; maternal, neonatal, and nutritional diseases

47%

fewer deaths of **children under 5** occurred in 2012 than in 1990

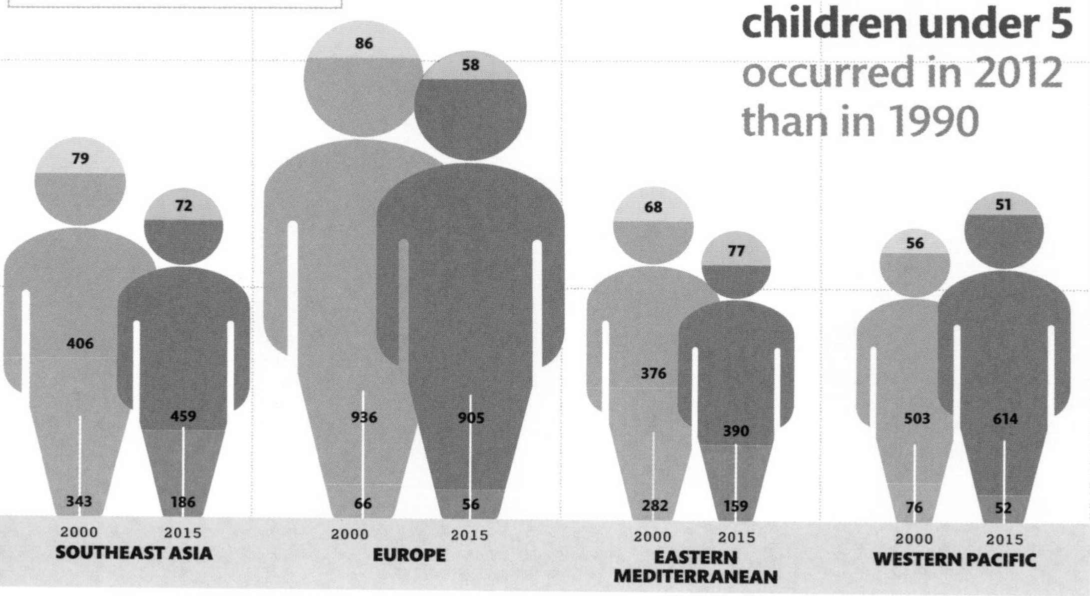

79
406
343
2000

72
459
186
2015

SOUTHEAST ASIA

86
936
66
2000

58
905
56
2015

EUROPE

68
376
282
2000

77
390
159
2015

EASTERN MEDITERRANEAN

56
503
76
2000

51
614
52
2015

WESTERN PACIFIC

Unequal world

Many people in the world are enjoying better lives, but inequality has grown dramatically. Disparities in wealth and income are seen both internationally and within individual countries.

Wealth inequality between countries can be demonstrated by looking at a country's gross domestic product (GDP) per person – a measure that gives a rough idea of income and standard of living. Rich countries such as Sweden are vastly better off than less developed nations such as Lesotho or Botswana.

Inequality also exists at a domestic level, which is quantified using the Gini coefficient, a statistical tool that measures differences in income. Recent economic growth in developed countries has mainly benefitted those at the top of society, widening the gap between rich and poor – a situation that is bad for everyone. Research shows that the more unequal a society, the more social problems it faces. Issues such as violent crime, mental illness, drug abuse, and teenage pregnancy are reduced in societies that are more equal.

Global inequality

Using both Gini ranks and GDP per capita shows that the most equal societies are also the richest. The world's most equal country, Sweden, had the sixth largest GDP per capita, while Lesotho, the least equal, had just US$996 GDP per head.

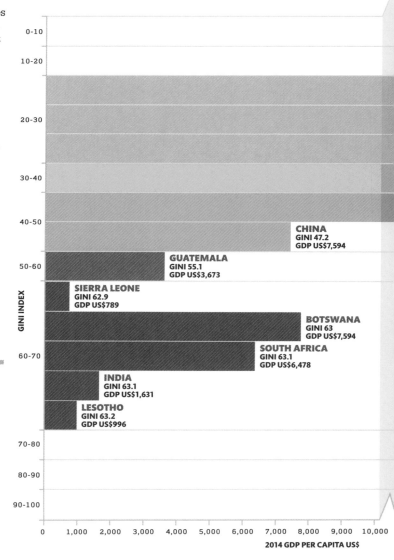

1% of the world's population has **more money than the other 99%** in 2016

SWEDEN
GINI 23.0
GDP US$58,887

SLOVENIA
GINI 23.7
GDP US$23,963

DENMARK
GINI 24.8
GDP US$60,634

UK
GINI 32.3
GDP US$45,603

USA
GINI 45
GDP US$54,630

What wealth is worth

Billionaires control around 10 per cent of the world's assets, yet many call poor nations home. A third of India's people live in poverty, but India ranks in the top five nations with the most billionaires.

KEY
National Gross Domestic Product (GDP)
Billionaires' net worth as % of national GDP

USA
536 BILLIONAIRES

CHINA
213 BILLIONAIRES

RUSSIA
88 BILLIONAIRES

INDIA
90 BILLIONAIRES

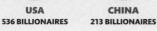

20,000 30,000 40,000 50,000 60,000

WHAT'S THE GINI COEFFICIENT?

Developed in 1912 by Italian statistician and sociologist Corrado Gini (1884–1965), the Gini coefficient is a measure of national equality, calculated by measuring how evenly income is distributed across a country. A country with perfect income equality will have a Gini coefficient of 0, while 100 indicates complete income inequality.

What a high Gini score means
Perfect inequality means one person has all the wealth while all others have nothing. In unequal nations, a few are very rich and a large number have very little.

What a low Gini score means
Perfect wealth equality would see all people having exactly the same amount of money, so countries with low Gini scores have more equal wealth distribution.

Corruption

In many countries efforts to combat poverty and halt environmental degradation have been seriously hampered by the effects of corruption. Corrupt practices often hit the poorest hardest.

Corrupt practices divert financial resources away from poor people and undermine controls intended to protect environmental assets such as forests and rare wildlife. Corrupt practices embrace a wide range of activities, including bribery, the embezzlement of public funds, obstruction of justice, and the concealment and laundering of the proceeds of corruption.

All of this can have disastrous impacts on economic development, as income inequalities increase, social policies are undermined, and economic growth stalls. In many countries suffering the effects of corruption, the exploitation of natural resources that should lead to overall development benefits instead enriches small elites. These conditions can contribute to civil war, as was the case in Sierra Leone in 1991.

Where corruption corrodes progress

According to the World Bank, each year corrupt practices lead to the siphoning off of about one trillion dollars. Funds desperately needed for education, healthcare, and other public services are thus lost, trapping people in poverty.

All sectors are affected, but water and power are particularly vulnerable to corruption because of the large number of public and business organizations involved in their supply. Corruption also leads to the flouting of laws that have been put in place to protect natural resources and ecosystems, leading to large-scale environmental damage. Protected wildlife species are traded on the back of bribes to customs officials while illegally logged timber enters international markets with forged papers.

Givers

Bribery can help commercial interests gain access to natural resources such as protected forests or fish stocks and is vital in getting illegally harvested goods to market. Businesses offer bribes in order to win public contracts. Bribes are paid to customs officials to turn a blind eye to the export or import of contraband, such as in the trafficking of ivory between Tanzania and China.

Water supplies

Bribes pay for licences to dispose of waste in open water, while large agribusinesses pay officials for access to irrigation.

- Corruption adds 30 to 45 per cent to connection costs of a clean water supply.

Essential services

Drugs intended for poor people are diverted for sale via private pharmacies. In addition, stolen funds hamper efforts to combat major health challenges, such as malaria and HIV/AIDS.

- The World Bank estimates that up to 80 per cent of non-salary health funds never reach some nations' local facilities.

What can we do?

- **Governments can bar companies** involved in corruption from bidding for public contracts.
- **Public bodies can instil** a zero-tolerance culture for corrupt practices.
- **Governments can prioritize** implementing UN anti-corruption policies.

Illegal animal trade

An unprecedented rise in illegal wildlife trading threatens decades of conservation work, making this the fourth most lucrative transnational crime, after drugs, arms, and human trafficking, worth between US$10 and $20 billion a year.

- At least 20,000 elephants are illegally killed for their tusks in Africa each year.

Forestry and illegal logging

Illegal logging now accounts for up to 30 per cent of the international timber trade. Cutting and shipping logs on the black market is a complex process and can only occur with the aid of corruption.

- The World Bank estimates that each year up to US$23 billion worth of wood is illegally cut, losing US$10 billion in revenue.

Takers

In all parts of the world, officials and politicians at all levels have been shown to be susceptible to taking bribes. In much of sub-Saharan Africa, for example, low salaries paid to public servants mean bribery is an open and accepted part of business. Such embedded corruption makes it extremely hard for many companies to conduct business legally.

Displaced people

The number of refugees, asylum seekers, and people displaced inside their own countries has rocketed. Forced out by war, persecution, and environmental change, the total is nearly that of Germany's population.

Following several years of substantial increases, the United Nations High Commission for Refugees estimated that, in 2019, the global total number of displaced people reached a staggering 7.5 million – about double the total in 2010. This forced movement on an unprecedented scale has created a "nation of the displaced". Its population includes refugees, asylum-seekers, and internally displaced people living within their home country. The causes include armed conflict, human rights violations, political violence, and the effects of drought. In 2019, the countries receiving the most people escaping across borders were Turkey, Colombia, Pakistan, Uganda, and Germany. For many host countries the additional demands arising from people seeking safety outside their their country of origin led to increased stresses on their already inadequate services.

A growing problem

By 2000, rapid globalization and the end of the Cold War had created new pressures that forced people to move, including those sparked by organized crime networks. In 2007, states with the most internally displaced people included Eritrea, Colombia, Iraq, and the Democratic Republic of Congo – all fuelled by internal conflict. Recent increases are largely due to the war in Syria, political divisions in Venezuela, and ongoing conflict in Afghanistan.

Somali refugee camp
People displaced from their homes have to find shelter in camps, often creating a huge strain on local resources.

KEY
- Internally displaced
- Refugees and asylum-seekers

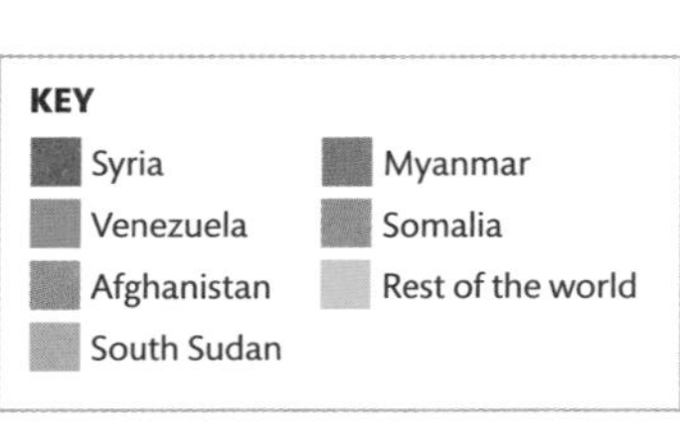

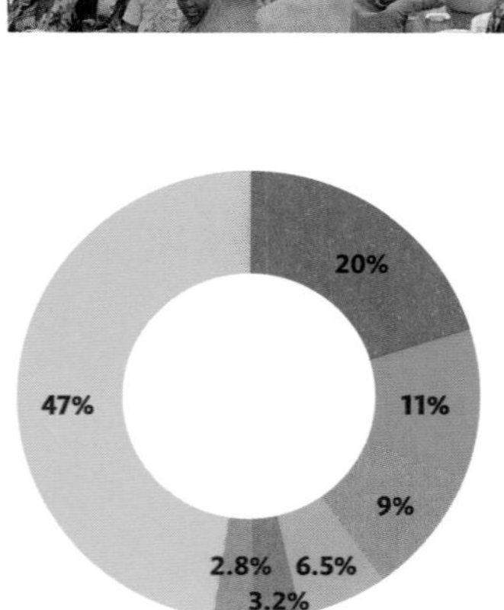

Where they come from
In 2019, over half the millions of refugees crossing international borders came from six countries: Syria, Venezuela, Afghanistan, South Sudan, Myanmar, and Somalia.

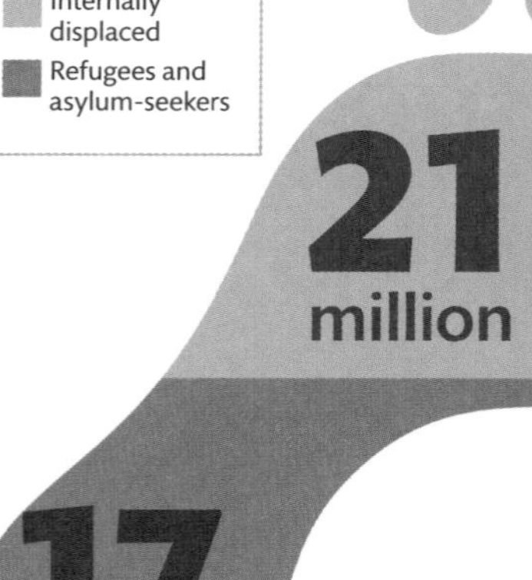

85%

The proportion of **forcibly displaced** people hosted **in developing countries**

46 million

34 million

2019

17 million

26 million

2007

HOW OLD IS A REFUGEE?

In 2014, just over half of all refugees were under 18 years old, up from 41 per cent in 2009. The proportion of people over 60 years old was far smaller. Although still a massive problem, in 2019, the proportion of internationally displaced people under 18 had dropped to 38 per cent.

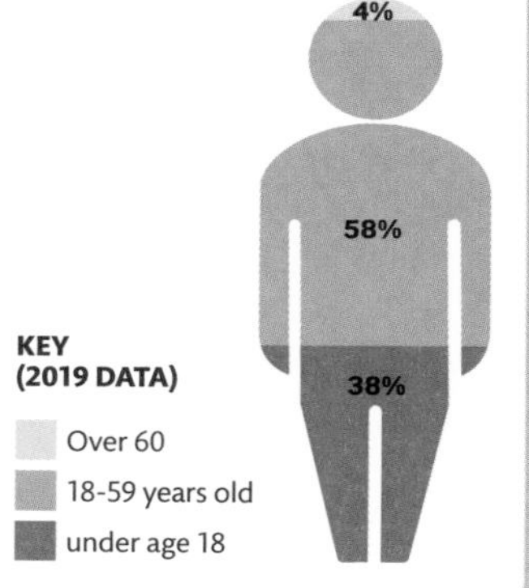

Our changing atmosphere

Without the atmosphere there could be no life on Earth. The shallow layer of gases that envelops our planet allows us to breathe and creates the climatic conditions we experience. Over the course of Earth's long history, the climate has changed many times. Natural factors caused this, but the main reason for recent climate change is the build-up of heat-trapping greenhouse gases produced by human activities (see pp.120–21). Because of this, the atmosphere is trapping more of the Sun's energy, causing average temperatures to rise, in turn altering the climate.

Carbon acceleration

The greenhouse gas most responsible for the recent warming of the atmosphere is carbon dioxide (CO_2). This trace gas occurs naturally and keeps Earth warm, maintaining favourable conditions for living things. The concentration of CO_2 fluctuates but has recently risen at an accelerating rate and is at the highest level it has been for at least 800,000 years. The main cause for this is the burning of fossil fuels, with some contribution from deforestation and emissions from soils.

Historical CO_2 levels
For thousands of years, carbon dioxide levels in the atmosphere remained below 280 parts per million (PPM).

400
350
300
250
PARTS PER MILLION (PPM)
10,000
7,500
5,000
2,500
0
YEARS BEFORE PRESENT DAY

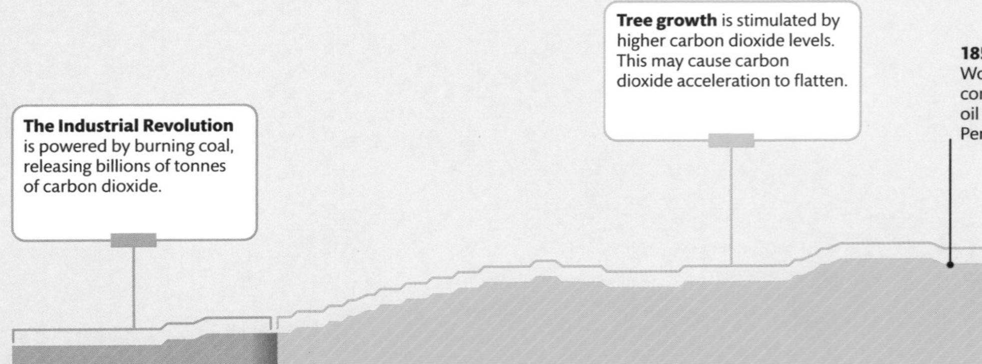

Tree growth is stimulated by higher carbon dioxide levels. This may cause carbon dioxide acceleration to flatten.
1859
World's first commercial oil field opens in Pennsylvania, USA
The Industrial Revolution is powered by burning coal, releasing billions of tonnes of carbon dioxide.
1750
1760
1780
1800
1820
1840
1860
18
YEAR

WHAT IS THE ATMOSPHERE?

The thin shell of gases that clings to the Earth is mostly comprised of nitrogen and oxygen. A handful of minor gases make up just one per cent. These include greenhouse gases, which trap outgoing heat from Earth's surface, keeping the surface and lower atmosphere warmer than it would be otherwise. The atmosphere contains the air that we breathe, protects us from the sun's radiation, and plays a vital role in the water cycle, and in climate patterns.

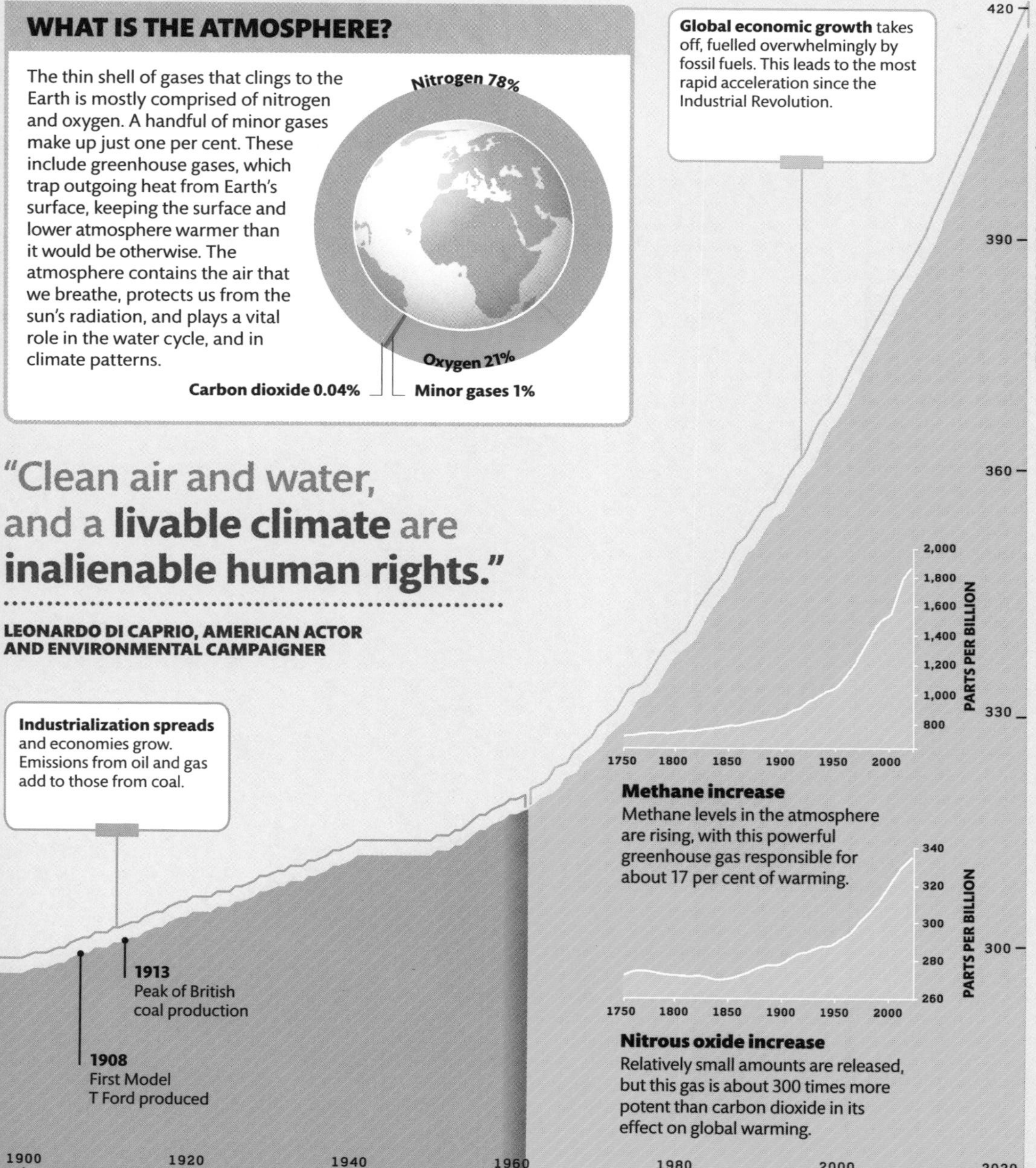

"Clean air and water, and a **livable climate** are **inalienable human rights.**"

LEONARDO DI CAPRIO, AMERICAN ACTOR AND ENVIRONMENTAL CAMPAIGNER

Methane increase

Methane levels in the atmosphere are rising, with this powerful greenhouse gas responsible for about 17 per cent of warming.

Nitrous oxide increase

Relatively small amounts are released, but this gas is about 300 times more potent than carbon dioxide in its effect on global warming.

The greenhouse effect

Light energy from the Sun is absorbed by Earth's surface, warming it up. The resulting heat is emitted from land and sea in the form of infrared radiation, most of which escapes back into space. However, heat-trapping gases in the atmosphere keep Earth much warmer than it would be otherwise. These gases create a "greenhouse effect", forming a layer that traps outgoing heat from Earth's surface and retains some of it within the lower atmosphere. Human activities interfered with Earth's delicate energy balance by rapidly increasing the concentration of greenhouse gases, causing the atmosphere to warm up.

Sources of greenhouse gases
Human activities produce greenhouse gases in many ways, but particularly by industrial activity and energy production.

EARTH'S ATMOSPHERE

Smaller amount of escaping infrared radiation

More trapped infrared radiation

4 Human activity causes an increase in the level of greenhouse gases.

5 More greenhouse gases prevent more heat from Earth's surface from escaping into space, raising the temperature of Earth's surface.

Industrial world
Industrialization has dramatically increased greenhouse gas concentrations, trapping more heat within the atmosphere and warming the surface and lower atmosphere.

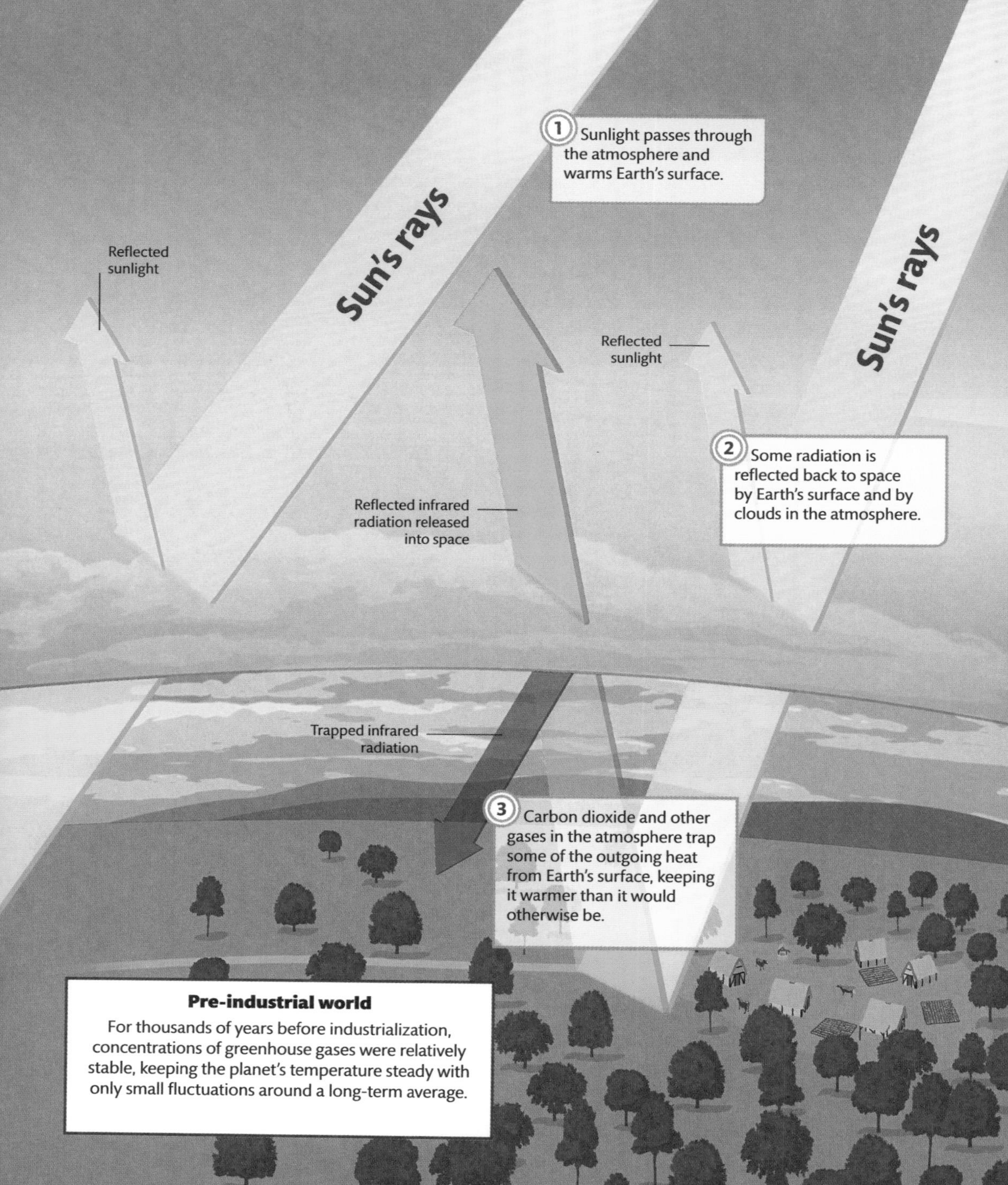
1 Sunlight passes through the atmosphere and warms Earth's surface.
Sun's rays
Reflected sunlight
Sun's rays
Reflected sunlight
2 Some radiation is reflected back to space by Earth's surface and by clouds in the atmosphere.
Reflected infrared radiation released into space
Trapped infrared radiation
3 Carbon dioxide and other gases in the atmosphere trap some of the outgoing heat from Earth's surface, keeping it warmer than it would otherwise be.
Pre-industrial world
For thousands of years before industrialization, concentrations of greenhouse gases were relatively stable, keeping the planet's temperature steady with only small fluctuations around a long-term average.

Hole in the sky

High in Earth's upper atmosphere, miles above the planet's surface, is a diffuse layer of ozone gas. Its presence protects life on Earth and is vital for the functioning of the planet itself.

Ozone formation relies on the oxygen in our atmosphere. As ultraviolet (UV) light from the Sun hits oxygen molecules in the stratosphere, ozone is formed and this in turn absorbs UV radiation that would otherwise damage the DNA (genetic material) of plants and animals. Oxygen was scarce until about 2.3 billion years ago, when an event called the Great Oxygenation occurred, the result of an increase in photosynthesis by microscopic organisms called cyanobacteria.

The ozone layer

Stratospheric ozone is highest between 20–30 km (12–19 miles) above Earth's surface, where the atmosphere is about a thousand times thinner than at ground level. Compounds released by human activities have depleted the ozone layer, raising concerns about greater levels of UV radiation reaching the surface. As well as damaging key groups of organisms such as marine plankton, increased UV exposure increases the risk of skin cancer in humans.

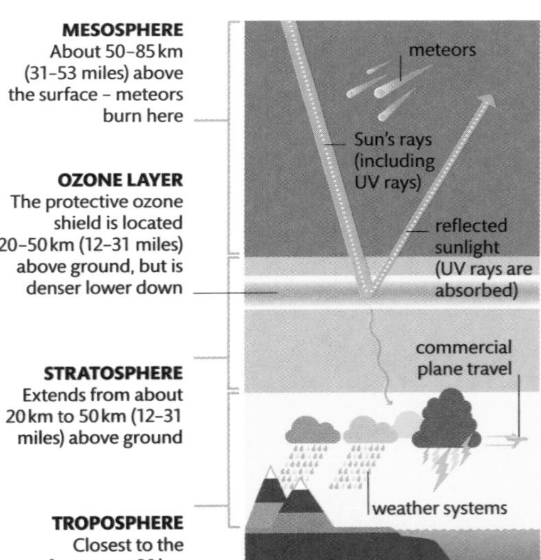

MESOSPHERE About 50–85 km (31–53 miles) above the surface – meteors burn here

OZONE LAYER The protective ozone shield is located 20–50 km (12–31 miles) above ground, but is denser lower down

STRATOSPHERE Extends from about 20 km to 50 km (12–31 miles) above ground

TROPOSPHERE Closest to the surface, up to 20 km (12 miles) thick

Antarctic ozone

Ozone concentration is measured in Dobson units (DUs). Prior to 1979, ozone had never been recorded below 220 DUs, but from then on it became apparent that during spring over Antarctica Earth's natural sunscreen was getting thinner. This area of depleted ozone became known as the ozone hole. In 1994 concentrations fell to just 73 DU.

THINNER OZONE Ozone-depleting substances are stronger at low temperatures, which is why the biggest depletion is over Antarctica

KEY

110 220 330 440 550

OZONE CONCENTRATION (DOBSON UNITS)

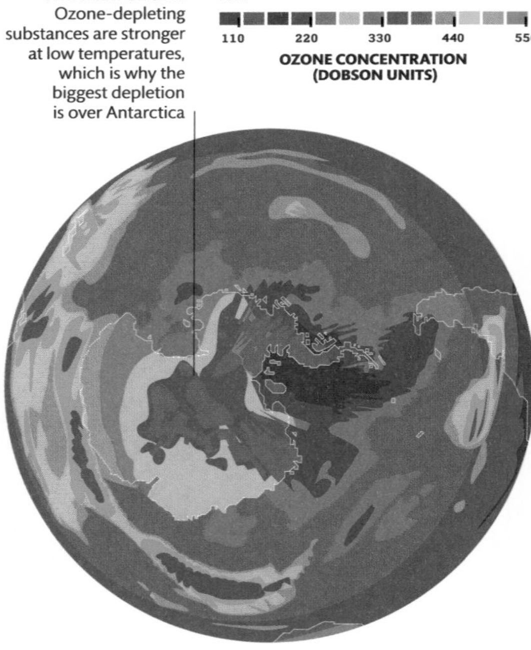

1979
Ground-based measurements of ozone began in 1956, at Halley Bay, Antarctica. Satellite monitoring started in the early 1970s and the first worldwide measurements began in 1978 with the Nimbus-7 satellite. The findings of this monitoring helped galvanize global political action.

NEW ZEALAND
From time to time the ozone hole breaks apart, and finger-like areas of depleted ozone extend over inhabited areas, including New Zealand

SOUTH AMERICA
During September 2015 the ozone hole spread over Punta Arenas, Chile, exposing inhabitants to very intense UV radiation

2020
The ozone hole was still huge in 2020, despite the phasing out of most ozone-depleting substances. In early October that year the hole covered about 25 million sq. km (9.6 million sq. miles), comparable to massive holes that formed in 2015 and 2018.

40%
depletion in ozone
above the Arctic in 2011

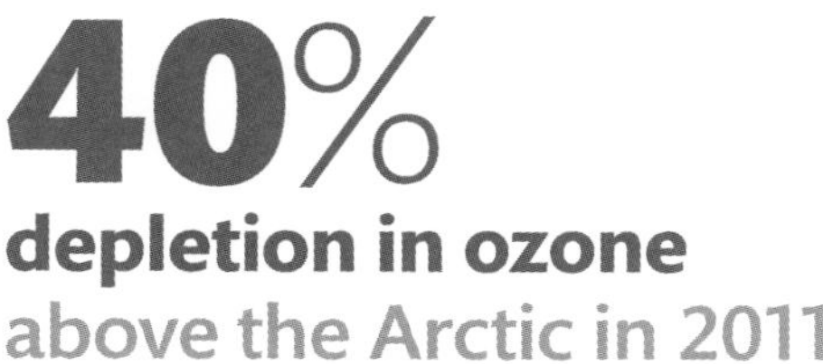

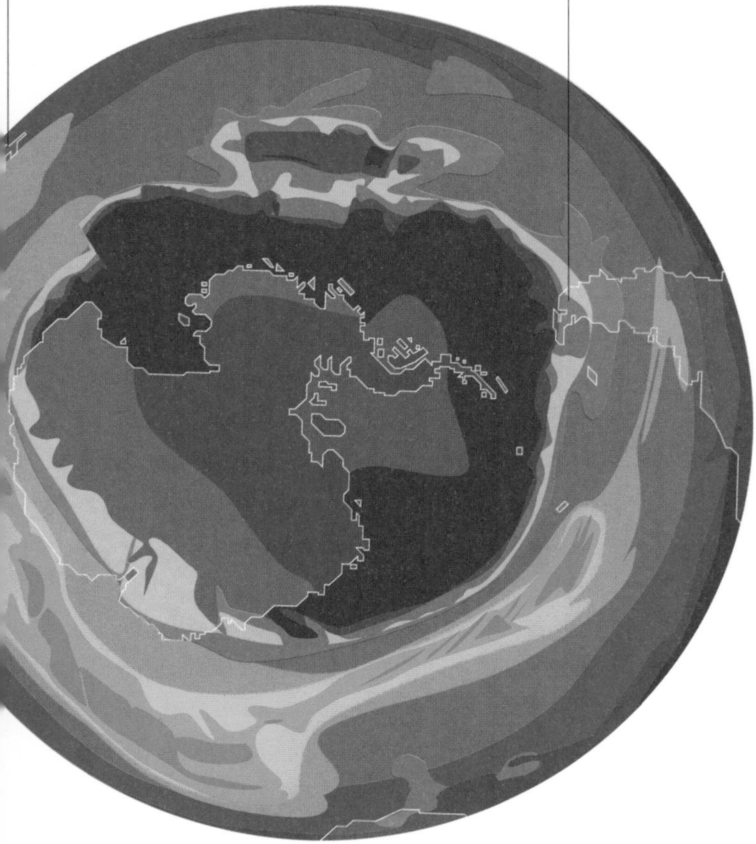

OZONE ENEMIES

When it became clear that certain chemicals were depleting the ozone layer, an international agreement – the Montreal Protocol – was negotiated in 1987. This successfully reduced the manufacture and release of ozone-depleting substances. Even so, ozone concentrations need time to recover. Meanwhile, monitoring ensures that warnings can be issued to areas at risk. Despite industry concerns about costs, alternatives to ozone-depleting substances were developed and are now widely used.

CFCs

Chlorinated fluorocarbons (CFCs) were used in aerosols, sterilization equipment, and refrigerators and freezers. Hydrofluorocarbons (HFCs) were used as substitutes.

Halons

These powerful greenhouse gases were used in fire extinguishers and technology systems employed by the aviation and defence industries. Production of halons ceased in 1994 under the US Clean Air Act.

Methyl bromide

Methyl bromide was used to control a huge range of agricultural pests. Many chemical and non-chemical alternatives now exist.

A warmer world

Rising temperatures, higher sea levels, and polar ice melt are just some of the many changes resulting from mankind's impacts on the atmosphere. These and other effects caused by increased concentrations of greenhouse gases are leading to a range of economic, social, and environmental consequences.

Flooding
Rising water levels already affect life in Bangladesh. The problem is likely to get worse.

The world is getting hotter. From 1850 to the present day, surface temperatures have risen by about 1°C (1.8°F) on average across the globe. The primary cause is undoubtedly the higher levels of heat-trapping gases, such as carbon dioxide, CO_2 (see pp112–13). This rise in temperature is already leading to melting ice sheets and glaciers, which in turn contributes to sea level rise. These changes are set to continue, but they may not be linear in relation to temperature increase. The planet's total amount of ice melt, for example, may accelerate as critical "tipping points" are reached – as is possible for the Greenland ice sheet and some Antarctic ice sheets.

10 million people each year are affected by coastal flooding

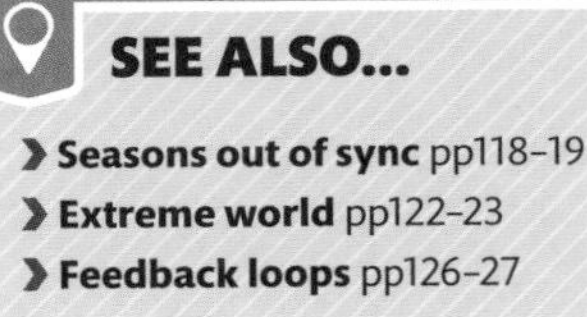

SEE ALSO...

› **Seasons out of sync** pp118–19
› **Extreme world** pp122–23
› **Feedback loops** pp126–27

Temperature rise

Throughout the northern hemisphere, 1983 to 2012 was probably the warmest 30-year period in the last 1,400 years. This map shows estimated global surface temperature changes from 1901 to 2012. Temperature decreases appear as shades of blue, while increases appear as shades of orange and purple. Areas with insufficient data are white.

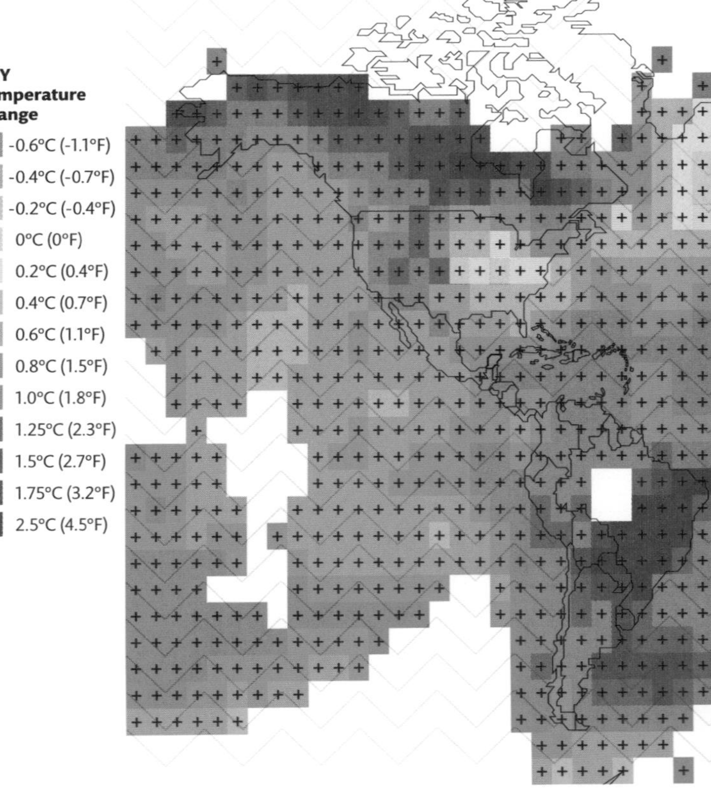

Rising waters

Sea levels are rising because of land-based ice melt and because ocean water expands as it gets warmer. The rate of sea level rise since the mid-19th century has been larger than the average rate during the previous two millennia. From 1880 to 2013, global mean sea level rose by about 23 cm (9 in). It will rise further as the ocean continues to warm and the melting of glaciers and polar ice sheets increases. The consequences of sea level rise are particularly severe in low-lying countries such as Bangladesh.

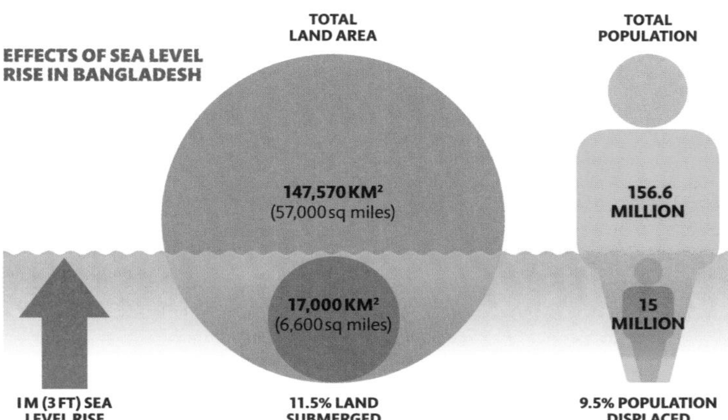

ICE MELT

The world has seen a massive loss of ice over the last two decades, from both ice sheets and glaciers. The average rate of ice loss from the Greenland ice sheet increased substantially from 2002 to 2011, and recent major ice loss has also been reported from Antarctica. The diagram below shows the seasonal shrinking of Arctic ice cover since 1970. By 2030, Arctic sea ice cover will be a fraction of 1970 levels. By 2100, there is likely to be little or no summer sea ice remaining here.

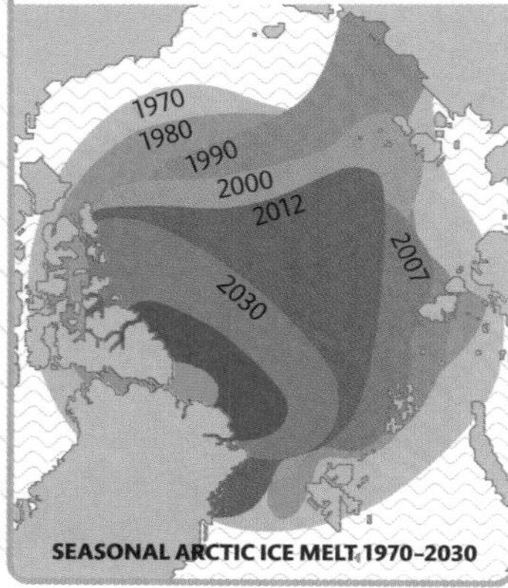

SEASONAL ARCTIC ICE MELT 1970–2030

Seasons out of sync

Across the world, climate change is leading to shifts in seasonal patterns. Sometimes subtle and taking place over decades, the implications can nonetheless be profound, for people and nature.

Many parts of the world have marked seasons that are important for farming, water supply, energy demand, and for sustaining the complex relationships between different wildlife species. Although many seasonal changes have been fairly predictable, longer-term shifts in climate are causing some patterns and relationships to fall out of balance – for example because of the earlier arrival of spring warmth and earlier flowering of plants.

Records reaching back decades, and in some cases centuries, allow scientists to document long-term trends. These records include data on the first and last leaves on ginkgo trees in Japan, the dates of first butterfly appearances in the UK, bird migrations in Australia, and of course temperature records that reveal increasingly short winters and the earlier arrival of spring. More important than these individual changes, however, is the impact that they may have on the many different and complex relationships between elements of the natural world.

SEE ALSO...

- **Farmed planet** pp58–59
- **Extreme world** pp122–23
- **How climate patterns work** pp120–21

Global impact

The natural world and the human civilizations that depend on it are heavily influenced by seasonal cycles. These cycles have been relatively stable and predictable for thousands of years. That is now subject to ongoing change, however, as the timing and intensity of temperature change and rainfall respond to global warming, impacting on people and wildlife in a variety of ways.

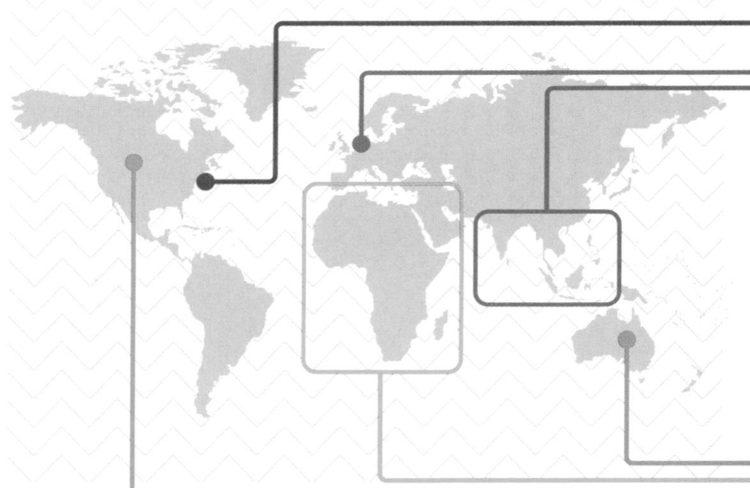

Earlier spring

Spring is arriving earlier across most of the USA. This map estimates the first day that leaves emerge in each state, comparing the average for 1991–2010 with the 1961–80 average. Such changes could have potential effects on plant and animal life cycles, which are tied to the seasons.

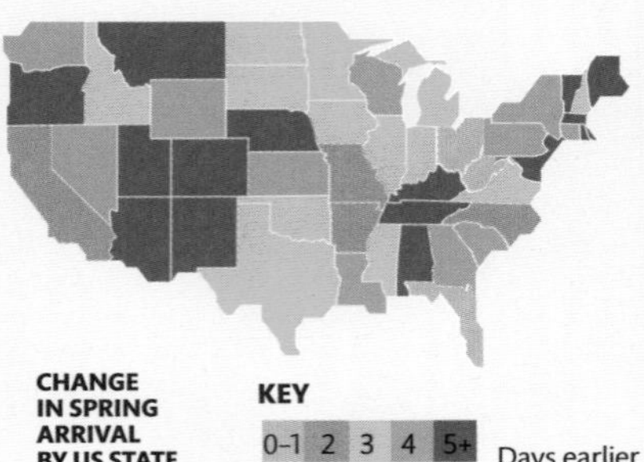

Warming waters

From 1982–2006 the North Atlantic Ocean warmed by about 0.23°C (0.4°F) per decade. Surveys running from the 1960s reveal that the commercially important summer flounder population has moved north, causing problems for fishing fleets.

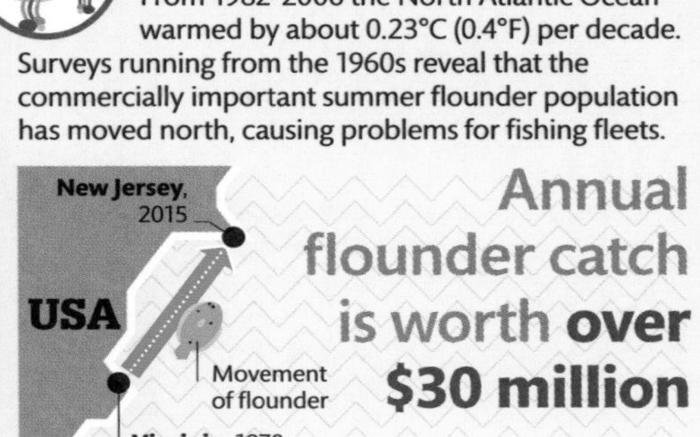

Annual flounder catch is worth **over $30 million**

Hungry birds

A Dutch study has shown how the breeding cycle of great tits has fallen out of sync with the peak abundance of caterpillars that the adult birds feed their chicks. Insects have adapted to earlier springs by breeding earlier, but the birds have not. This could lead to lower chick survival rates.

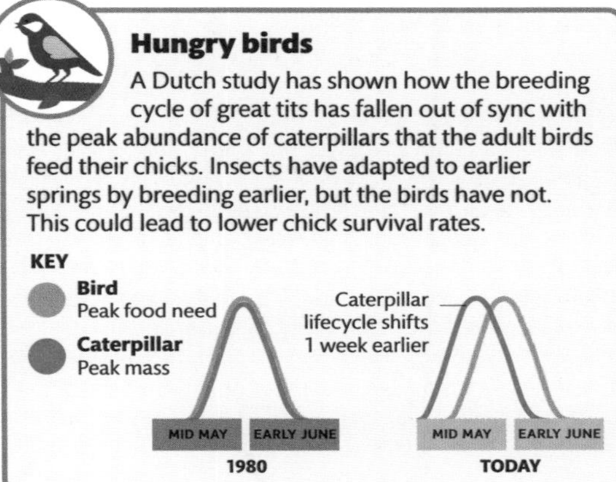

First leaves and blooms came **one day earlier per decade** in the Northern hemisphere from 1955 to 2002

Indian monsoon

The Indian monsoon is a stable, reasonably predictable annual weather pattern, but it is believed that rainfall each year will become more variable as the climate warms. Both flooding and drought (between the rains) are predicted to increase. Even a 10 per cent change can have huge impacts on farming, food prices, and the economy.

5–10% increase in monsoon rain

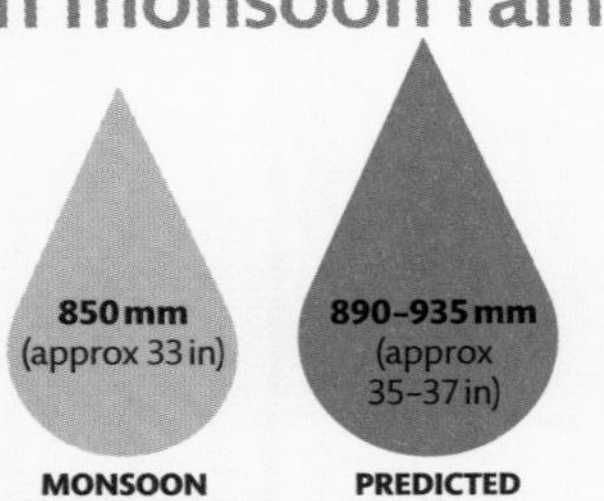

Farming

More than 70 per cent of African farmers rely on rain (rather than irrigation) to produce food. Changes in the timings and intensity of seasonal rains are leading to reduced yields and lower incomes.

Rainfall

Australia is the driest inhabited continent, and changes in average rainfall have a major impact on farming. Scientists believe the Australian climate has already changed with recent droughts revealing the cost of less rain. More intense heavy downpours have also affected some areas.

Warming up

Seven of Australia's 10 warmest years on record occurred in the 13 years from 2002, with a record mean temperature for 2005–2014. High temperatures worsen the effects of low rainfall.

Bushfires

The drying climate of southeastern Australia has increased the risk of bushfires. From 1973 to 2007 there was an overall increase in high-fire-danger weather.

How climate patterns work

Climate is determined by the interaction of a finely balanced set of factors. Solar energy warms the oceans and the atmosphere, while differences in atmospheric pressure and temperature drive air and sea currents. Climate is influenced by latitude as well as factors such as distance from the ocean and height above sea level. Climatic conditions are measured in averages over decades. Weather is shorter term, changing from day to day. Solar heating causes the air in the Earth's atmosphere to cycle around the globe in three sets of giant loops, called atmospheric cells – Hadley, Ferrel, and Polar cells. These produce north–south airflows, which are modified by Earth's spin, producing winds that blow diagonally.

Cold high-level air flows south

Cool air sinks in subtropical latitudes

Cold high-level tropical air flows north

Warm surface current

Cold, deepwater current

Ocean currents
Oceans absorb energy from the Sun and move it around in surface currents. Ocean currents carry warm tropical water into cooler regions, affecting the climate there.

Cold air is drawn north

SEASONS

Earth rotates around the Sun on a tilted axis in an orbit that takes one year. As different areas of the Earth face towards or away from the Sun, the length of day and temperature changes. This leads to long days and short nights in summer and the reverse in winter. Seasons are most pronounced near each pole.

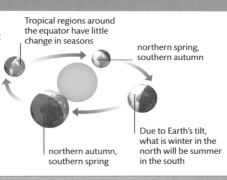

Tropical regions around the equator have little change in seasons

northern spring, southern autumn

Due to Earth's tilt, what is winter in the north will be summer in the south

northern autumn, southern spring

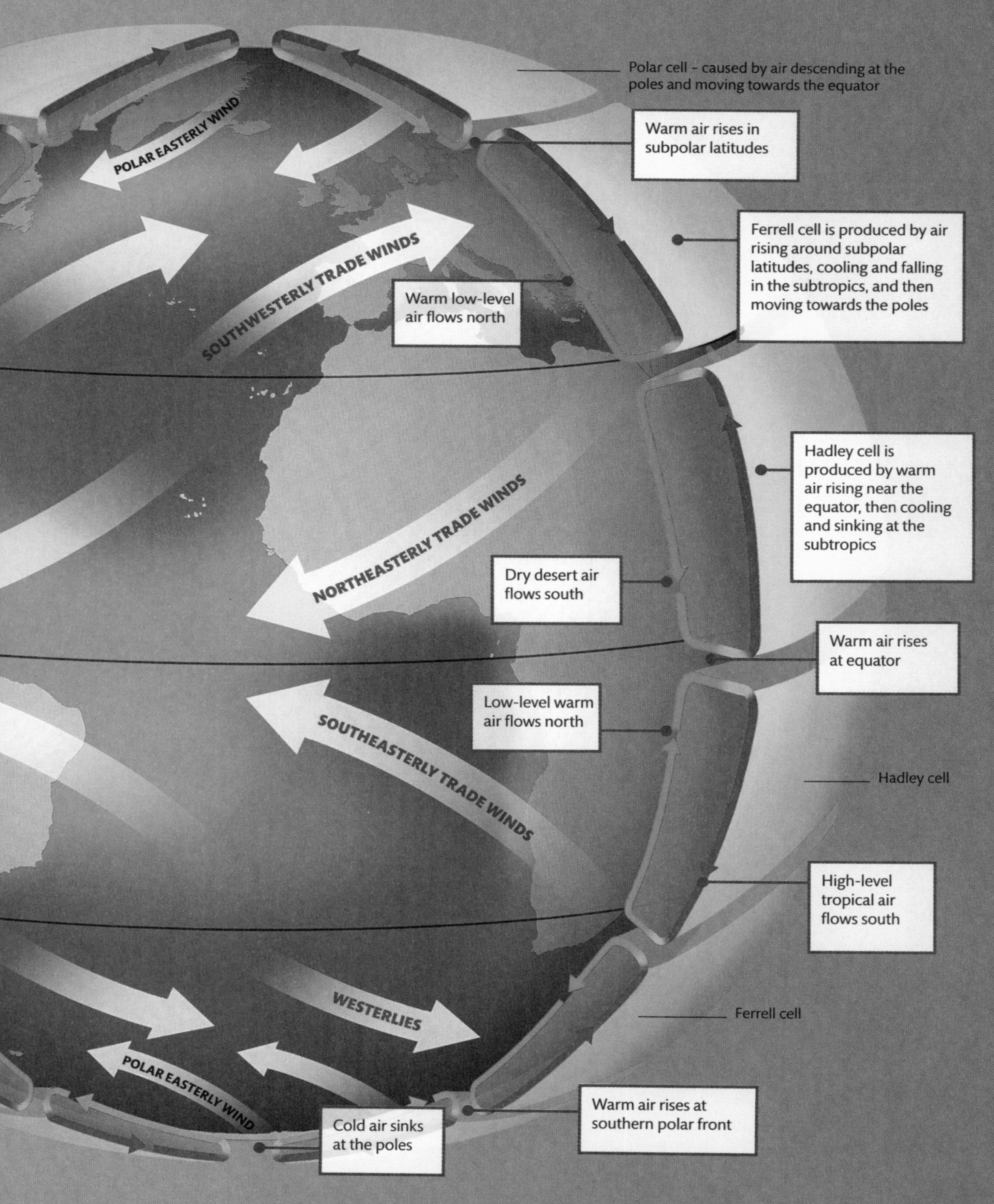
Polar cell - caused by air descending at the poles and moving towards the equator
Warm air rises in subpolar latitudes
POLAR EASTERLY WIND
SOUTHWESTERLY TRADE WINDS
Ferrell cell is produced by air rising around subpolar latitudes, cooling and falling in the subtropics, and then moving towards the poles
Warm low-level air flows north
Hadley cell is produced by warm air rising near the equator, then cooling and sinking at the subtropics
NORTHEASTERLY TRADE WINDS
Dry desert air flows south
Warm air rises at equator
Low-level warm air flows north
SOUTHEASTERLY TRADE WINDS
Hadley cell
High-level tropical air flows south
WESTERLIES
Ferrell cell
POLAR EASTERLY WIND
Warm air rises at southern polar front
Cold air sinks at the poles

Extreme world

Weather records are being broken around the world. As the climate warms, extreme weather is becoming more frequent and leading to a series of devastating knock-on consequences.

The build-up of more heat in the atmosphere is leading to changed patterns of evaporation and atmospheric circulation. This causes unusual and extreme weather. Weather is highly variable in the short term, but climate trends are based on averages that span decades. The trend towards more extreme weather events is in line with the predicted impacts of progressive warming. A continued increase in warming will lead to more extreme conditions, in turn generating a wide range of economic, social, and environmental consequences. These are being intensified by other environmental changes, including deforestation.

Weather warning

The impact of more extreme weather events will undermine food production, place increased pressure on emergency services, elevate demand for humanitarian assistance, create security tensions, and exacerbate conflict. A vital aspect of future economic planning will be to prepare for extreme events, so that their impact is reduced and a quick recovery can occur. This might be through storing more rainwater, conserving and restoring forests, adopting new standards for infrastructure, improving soil quality, and developing more diverse agriculture.

Hurricanes
The intensity, frequency, and duration of North Atlantic hurricanes, such as Hurricane Dean (pictured hitting the coast of Mexico in 2007), have all increased since the early 1980s.

Droughts

Australia, California, parts of East Africa, and southern Brazil have all recently suffered the effects of severe drought. This has resulted in limited availability of water for industry, farming, households, wildlife, and energy generation.

Floods

Devastating floods have recently affected parts of West Africa, Thailand, western Europe, and South America. This has led to loss of life, damage to property, and major interruptions to business activity. Damage to soil caused by farming has made flood events more severe.

Storms

As oceans heat up, storms powered by warm air rising from them are on average becoming more violent. The severest tropical cyclones ever recorded have occurred during the last decade. As the world warms severe storms are expected to become more frequent.

Food shortage

Floods, droughts, and storms can reduce food production. This causes shortages, rising prices, and hunger among the poorest people. Recently, drought and heatwaves have hit yields in the USA and Australia.

Lack of drinking water

Severe droughts have recently led to restrictions on public water consumption, including in parts of Australia, Brazil, and the USA. Flooding and storm damage can lead to contaminated drinking water supplies.

Homelessness

Huge floods, such as those in Pakistan, destroy thousands of homes. Over recent years, cyclones have devastated islands and coastal areas, leaving tens of thousands of people homeless.

Damaged infrastructure

Roads, ports, railways, and power-distribution systems are all affected by severe weather. This damage has increased the level of weather-related insurance claims.

Mass migration

Many of the migrants arriving in Europe over recent years have come from parts of Africa that are suffering from the effects of desertification, in turn made worse by reduced or more erratic rainfall. In the future others will be forced to move by rising sea levels. Many also flee from conflict, which may in turn be connected to the impacts of severe weather events.

Conflict

The impacts of severe weather can be linked to conflict. The Syrian Civil War began at a time of severe drought. This aggravated political tensions when about 1.5 million rural people were forced to move to urban areas. The exceptional dry period was consistent with climate change projections for the eastern Mediterranean that suggest progressively less rainfall there.

Human casualties

Some storm events cause massive loss of life directly, such as that brought about by Hurricane Mitch in 1998. About 18,000 people lost their lives as a result of this exceptional storm, which also wrecked infrastructure across large swathes of Central America. The impacts of extreme events – such as hunger, exposure, and conflict – also lead to human casualties.

The 1.5-degree limit

In 2015, at the Paris climate summit, governments agreed the need to keep global temperature rise to below 2°C (3.6°F) compared with pre-industrial times and to aim for the more challenging limit of 1.5°C (2.7°F).

Meeting the target to limit global warming to below 2°C (3.6°F), and aiming for no more than 1.5°C (2.7°F), is necessary to achieve the central aim of the original 1992 UN Framework Convention on Climate Change of preventing "dangerous" human interference with the climate system. Although there is no single scientific verdict as to what constitutes "dangerous", the 1.5°C (2.7°F) goal is needed to reduce the risk of disastrous impacts on food production (see pp68–69), water security (see pp72–73), ocean acidification (see pp152–53) as well as to avoid the ever-greater risk of triggering major climate system shifts (pp126–27). Based on the temperature limit, a "carbon budget" can be set. This is the amount of carbon that can be emitted before global temperature is likely to exceed the agreed level.

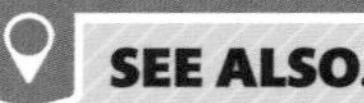

SEE ALSO...

- **A warmer world** pp116–17
- **How much can we burn?** pp128–29
- **The carbon crossroads** pp130–31
- **Targets for the future** pp134–35

Our carbon budget

The carbon budget sets a limit on human carbon emissions. For a 50:50 chance of limiting temperature increase to 1.5°C (2.7°F), there was an estimated 157 gigatonnes of carbon (GtC) remaining in the budget in 2017. For a two-thirds chance of meeting the target, the budget in 2017 was 114GtC. If the warming effect of other greenhouse gases is added, the budget is smaller. 2020 was the warmest year recorded, marking an average global temperature increase of more than 1°C (1.8°F), compared with pre-industrial times.

In 2019 the UK was the first major country to set a legally binding target committing to net zero carbon emissions by 2050

The carbon dioxide/ temperature link
If we continue to release CO_2 into the atmosphere at the present rate, by 2100 average global temperature increase will be between about 3°C (5.4°F) and 5°C (9°F), compared with the mid-19th century.

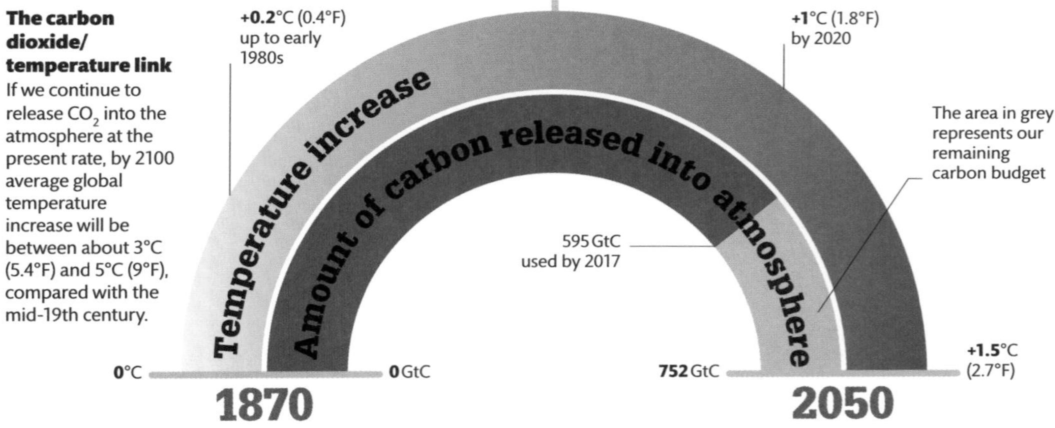

In 2017, we had 157 GtC left in the global carbon budget

2030

Carbon budget exhausted in 2031, based on annual emissions at 2017 level

2025

2020

The predicted remaining carbon budget in 2020 was about 270 GtC

2017

Time to act
In 2017, the remaining carbon budget was set to last about 14 years. The faster we "spend" the budget, the more quickly we commit the world to breaching the 1.5 degree limit.

Carbon budget in 1870 was 752 GtC

By 2014, we had used up 595 GtC

FINDING THE RIGHT PATH

A range of different strategies are needed to follow an emissions pathway consistent with a maximum 1.5°C (2.7°F) temperature increase. Many of these strategies relate to energy choices but also to deforestation, land use, and economic policies. Some measures are being taken, but much more action is urgently required to meet the agreed target.

Electricity efficiency Emissions can be reduced through efficient uses of power: for example by fitting modern electrical motors in factories or LED lightbulbs in homes.

Renewable electricity Switching from fossil fuels to renewable alternatives will be a major focus in getting onto an emissions pathway for two degrees.

Carbon capture Capturing waste CO_2 and storing it (see pp128–29) can reduce emissions from power stations, although limited progress is being made with this technology.

Vehicle efficiency A switch to electric vehicles and using hydrogen for trains, ships, and planes could reduce carbon emissions and make air cleaner to breathe.

Low-carbon fuels Sustainably produced bioenergy to generate power and replace liquid fuels could play a limited role in reaching carbon targets.

Smart growth Building communities with housing and sustainable transport options close to offices, schools, and shops would protect the environment and support local economies.

Carbon taxes Imposing a price on carbon emissions would send a clear economic signal and encourage investment towards cleaner energy sources and away from fossil fuels.

Forest and soil carbon Halting deforestation and restoring forests could make a significant contribution to meeting the two-degree target, as well as helping conserve wildlife and water.

Switching subsidies Removing fossil fuel subsidies could lead to around a 13 per cent cut in emissions. These valuable subsidies could then support sustainable alternatives.

Feedback loops

While reducing fossil fuel emissions and limiting land-use change is under some human control, so-called feedbacks play an increasingly important part in climate change as our world gets warmer.

Climate feedbacks are effects of climate change that either speed up (positive feedback) or slow down (negative feedback) warming. For example, certain types of cloud that might become more abundant at higher temperatures could create a cooling effect and slow down the speed of climate change. The warmer the world becomes, the greater the risk that major positive feedbacks will hasten climate change regardless of any actions taken to cut emissions.

The Amazon drought of 2010 caused the **release** of about **two billion tonnes of carbon**

Feedback loops and their impact

Rising global temperature could trigger serious positive feedbacks that risk the acceleration of climate change. This is one reason why, in 2015, governments adopted the goal of limiting warming to 1.5°C (2.7°F) above the average global temperature before 1850. Higher temperatures increase the likelihood of feedbacks, including loss of ice cover, melting permafrost, dieback of rainforests, and the release of methane from the seabed.

CO_2 released

Arctic melt

Most of the Sun's energy that hits icy surfaces is reflected back into space. As ice melts in the Arctic and elsewhere, the darker surfaces of ocean and tundra are exposed. These absorb much more of the Sun's energy, speeding up global warming, and, in turn, melting more ice.

Seabed methane release

Huge quantities of methane are stored in seabeds. This methane is stable at lower temperatures, but global warming could cause the gas to be released into the atmosphere. This powerful greenhouse gas would speed up warming and cause further methane release from the seabed and permafrost.

Permafrost melt

At high latitudes, close to polar regions, there are vast areas of peaty soils frozen as permafrost. They contain trapped carbon dioxide and methane. As the climate warms and causes the permafrost to melt, these greenhouse gases are released. As more of these gases are released, there is further melting and greater emissions.

Rainforest dieback

Decreased rainfall and heat stress could cause large areas of rainforest to dry out and turn into savanna, or grasslands. These ecosystems hold less carbon than dense forests, thereby increasing levels in the atmosphere. Changes to forests will also affect wildlife.

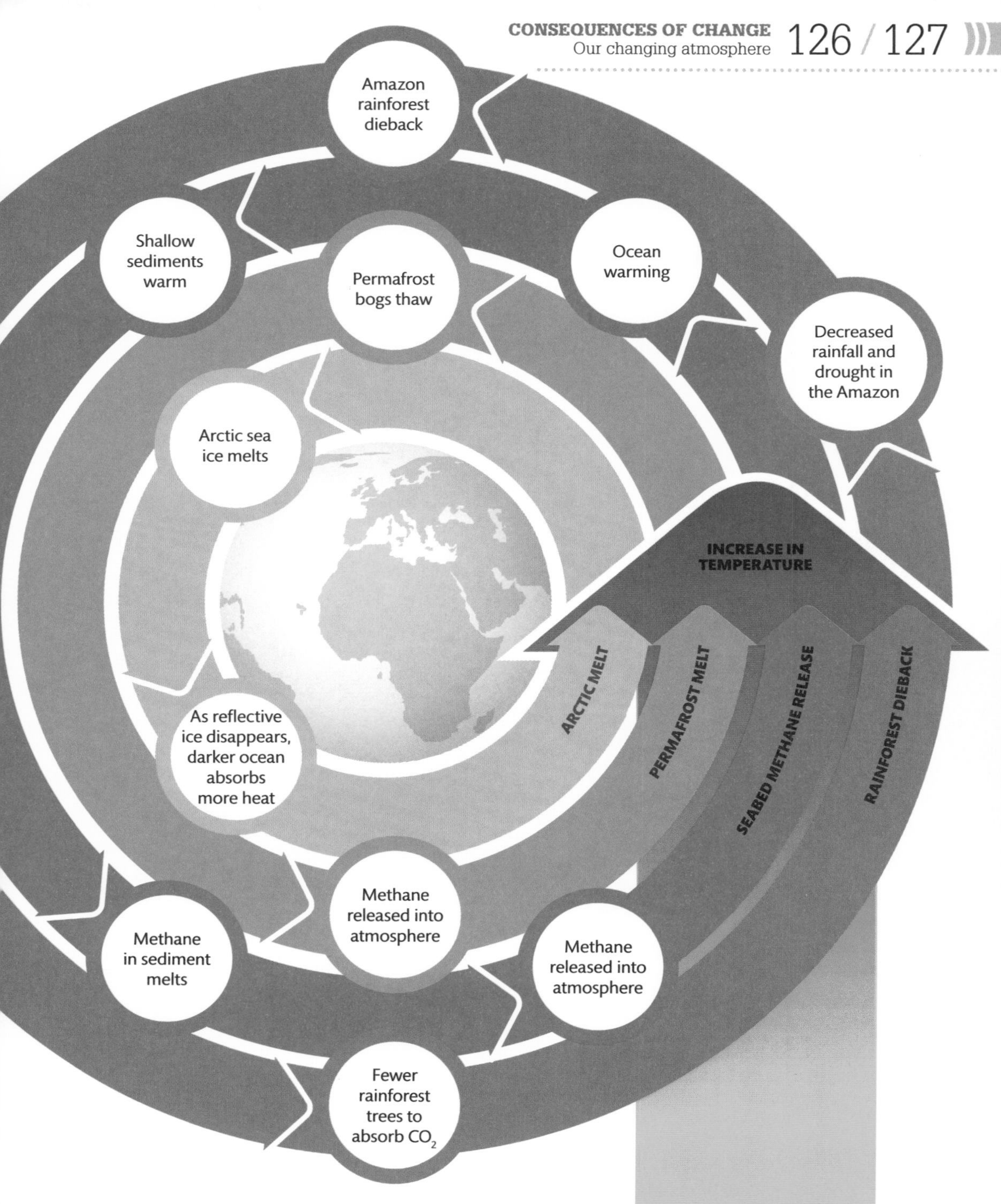
Amazon rainforest dieback
Shallow sediments warm
Ocean warming
Permafrost bogs thaw
Decreased rainfall and drought in the Amazon
Arctic sea ice melts
INCREASE IN TEMPERATURE
ARCTIC MELT
PERMAFROST MELT
SEABED METHANE RELEASE
RAINFOREST DIEBACK
As reflective ice disappears, darker ocean absorbs more heat
Methane released into atmosphere
Methane in sediment melts
Methane released into atmosphere
Fewer rainforest trees to absorb CO_2

How much can we burn?

It is now possible to calculate the amount of greenhouse gases that can be emitted before temperature thresholds are exceeded. With this in mind, we must decide how best to use the fossil fuel reserves we have.

Carbon budgets describe the amount of greenhouse gases, especially carbon dioxide (CO_2), that countries agree can be released into the atmosphere. These budgets are being compared with known fossil fuel reserves to determine the amount of coal, oil, and gas that can be burnt before the world commits to a dangerous temperature increase. A hazardous level of warming was agreed in 2015 to be 1.5°C (2.7°F) above the global average temperature in pre-industrial times (see pp124–25). It is estimated that to have a 50:50 chance of keeping the overall temperature increase to below 1.5°C (2.7°F), only about a fifth of known reserves can be burnt.

Staying on budget

There is a much larger amount of fossil fuels in the ground than we can safely burn. Potential emissions of CO_2 from total known reserves are calculated to be 762 GtC (gigatonnes of carbon). This excludes any new deposits that are yet to be discovered. Effective action on climate change would thus lead to coal, oil, and gas companies' assets being left in the ground or "stranded".

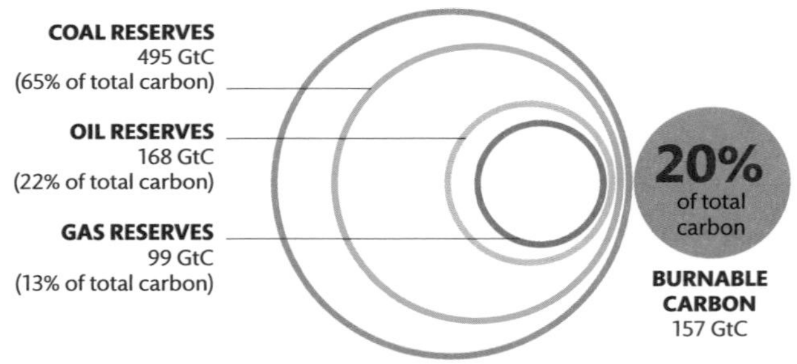

CARBON CAPTURE TECHNOLOGY

A process that traps carbon emissions at source and compresses the gas into liquid form that is stored in geological structures may allow some fossil fuels to be used without exceeding the 1.5°C (2.7°F) carbon budget. However, this technology is not widely used.

Unmineable coal seams
Carbon dioxide can be injected into deep, inaccessible, or otherwise uneconomic coal deposits for storage. During this process methane, a greenhouse gas, is released. The methane can then be recovered and used as an energy source.

Depleted oil deposits
Oil and gas fields that are nearing the end of their productive lives can be used for carbon storage. Injecting carbon dioxide can increase pressure in depleted oil fields to retrieve more oil in a process known as enhanced oil recovery.

Deep saline geology
Deep geological formations made up of sandstones and limestones that hold salty water are sometimes impermeable because they are capped by another type of formation. This means that they are able to hold injected carbon dioxide.

Fuel reserves
If the available carbon budget for 1.5°C (2.7°F) is allocated equally across the main fossil fuel reserves, it becomes clear that we can burn only a fraction of known reserves. This analysis has major implications for the future of fossil fuel companies, and the investors who presently derive income from their profits.

11% of coal is burnable

Coal reserve 495 GtC

52% of gas is burnable

Gas reserve 99 GtC

31% of oil is burnable

Oil reserve 168 GtC

28% of global carbon dioxide emissions in 2020 were emitted by China

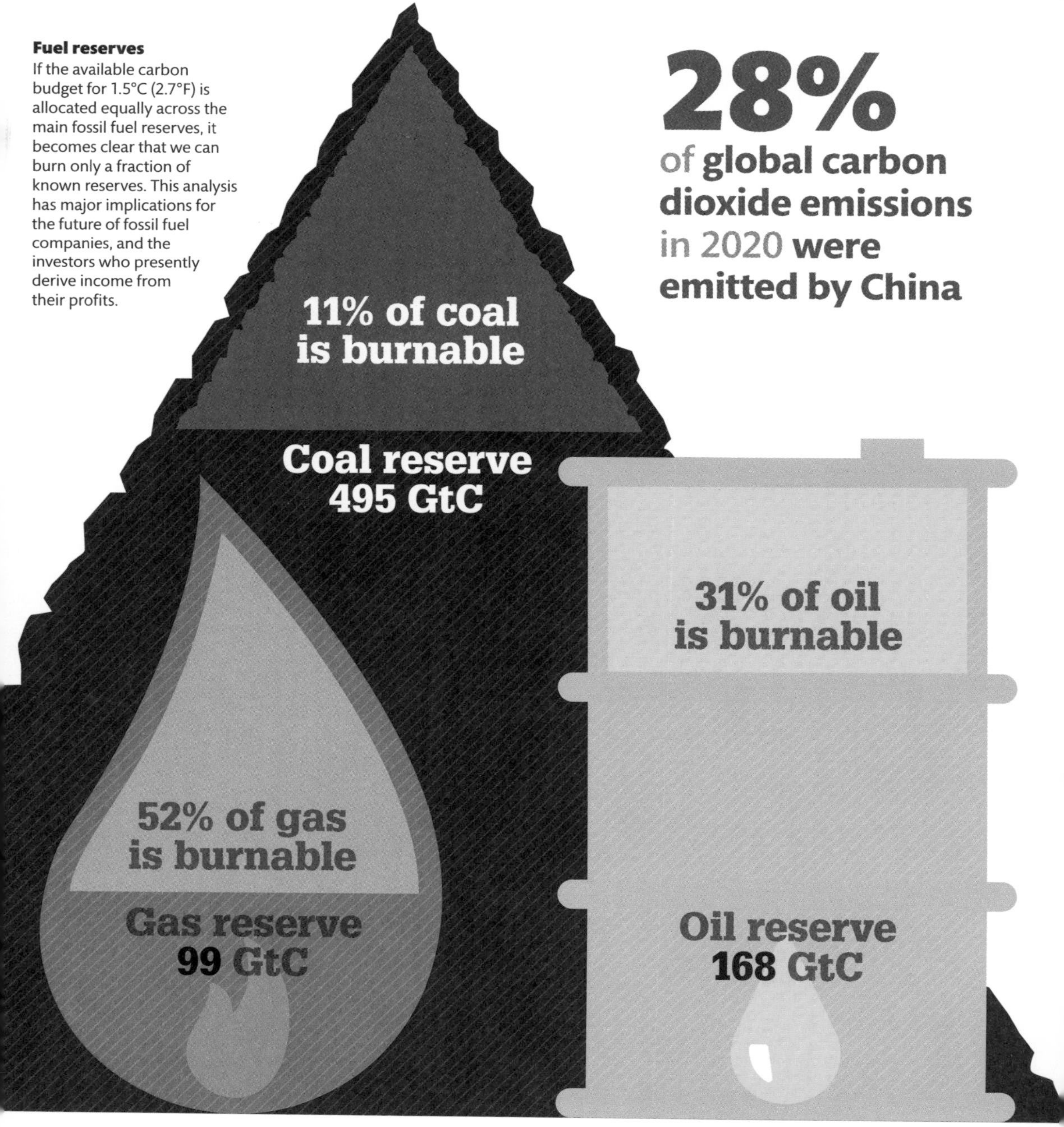

The carbon crossroads

The world is at a crossroads. To limit global warming to below two-degrees Celsius (3.6°F) above the pre-industrial average temperature, action is required now.

Future concentrations of carbon dioxide (CO_2) and other greenhouse gases in the atmosphere will be determined by a range of factors, including energy sources and population change. Without urgent action, it will be nearly impossible to limit global warming to below 2°C (3.6°F), still less to the target of 1.5°C (2.7°F) adopted in 2015.

SEE ALSO...

- **A warmer world** pp116–17
- **The 1.5-degree limit** pp124–25
- **How much can we burn?** pp128–29
- **Targets for the future** pp134–35
- **What's the global plan?** pp182–83

Past, present, and future

The Intergovernmental Panel on Climate Change's (IPCC) Fifth Assessment Report, which was finalized in 2014, is a comprehensive assessment of climate change. Among its key findings were that human activities, especially the release of CO_2, are causing a sustained, unequivocal rise in global temperatures. Even if all emissions are stopped immediately, temperatures will continue to rise as a result of greenhouse gases already in the atmosphere. Mitigating this rise will need significant and lasting reductions of greenhouse gas emissions from this point onwards.

WHAT IS AN RCP?

As part of their findings, the IPCC explored four contrasting scenarios of future climate change. These scenarios, known as Representative Concentration Pathways (RCPs), project greenhouse gas concentrations and their impact on global temperatures over the 21st century. Each pathway is consistent with different scenarios based on various socio-economic trends and policy choices.

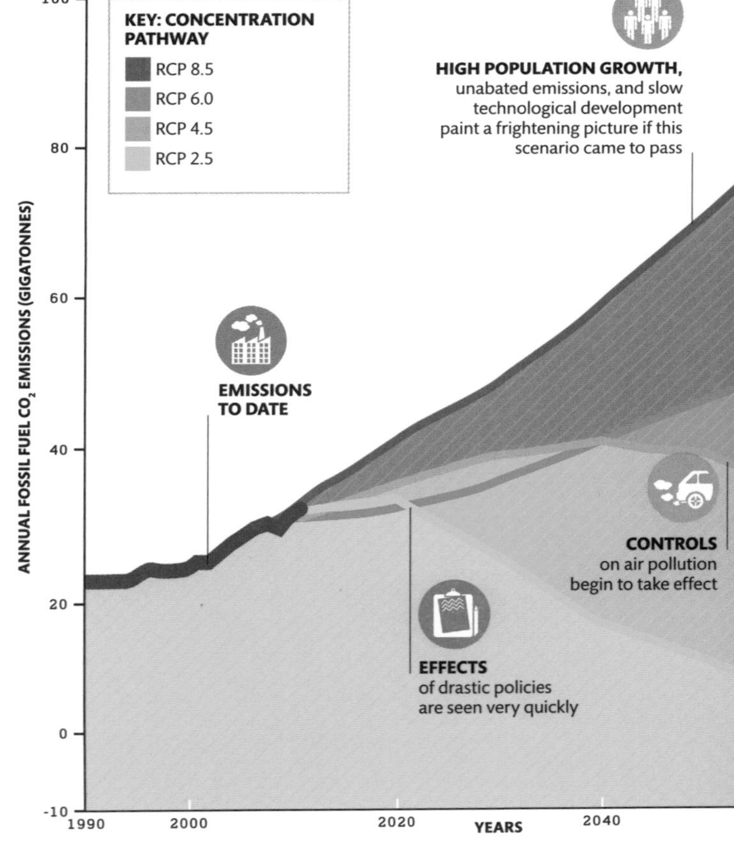

"We received this world...as **a loan from future generations**, to whom **we will have to return it!**"

POPE FRANCIS

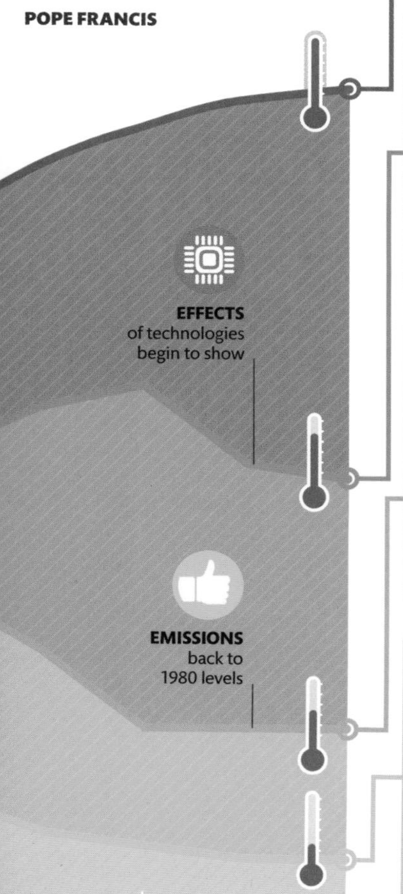

Highest concentration pathway

RCP 8.5 is consistent with high population growth, lower incomes in developing countries, slow technology development, and rising emissions from burning fossil fuels. Emissions eventually level out, but the average global temperature increases by around 5°C (9°F).

Ecosystem failure Many ecosystems, such as large areas of tropical rainforest, will collapse, releasing more CO_2.

High concentration pathway

RCP 6 is consistent with technological advances that begin to have a large-scale impact during the 2080s. This causes concentrations of CO_2 and other greenhouse gases to stabilize at around the year 2100. Under this scenario, the average global temperature increase is about 3°C (5.4°F).

Food shortages Changing rainfall and temperature reduce food production, especially in the tropics.

Medium concentration pathway

RCP 4.5 is consistent with moderate action on climate change and air pollution. Forest conservation and regrowth brings significant positive effects from the 2040s to 2060s. During the 2080s, emissions are about the same as during the 1980s. The increase in temperature is 2–3°C (3.6–5.4°F).

Reef loss About two-thirds of the world's coral reefs suffer major long-term degradation.

Low concentration pathway

RCP 2.5 is consistent with an early peak and then a decline in emissions, arising from a radical and almost immediate policy change to encourage renewable energy, energy efficiency, and forest conservation on a large scale. In this scenario, the overall average temperature stays below the critical 2°C (3.6°F) mark.

Declining milk production Lower-quality pasture and higher heat stress on cows affects major dairy exporters such as Australia.

The carbon cycle

Carbon is essential for life and is present in all living things. It flows in cycles through Earth's systems, passing between rocks, plants and animal life, the atmosphere, and the oceans, including as carbon dioxide (CO_2). This finds its way into the air via respiration and as a result of burning. It is taken out of the air mostly by photosynthesis (see p164) and absorption into seawater (see pp152–53). During the last two centuries human activities have seriously disrupted the carbon cycle, causing more CO_2 to build up in the atmosphere, mostly because of burning fossil fuels and deforestation. The graphic shows carbon circulating between different parts of the Earth system.

123 BILLION TONNES (135 BILLION TONS)

92 BILLION TONNES (102 BILLION TONS) **ABSORBED BY OCEANSC**

All plants, including trees, absorb CO_2 from the atmosphere, using it in photosynthesis.

When animals produce waste or die, they add dead matter (containing carbon) to the soil.

Oceans absorb CO_2 from the atmosphere. Some is used by phytoplankton in photosynthesis – or ends up as carbonate in shells of marine animals. But too much can contribute to acidification of the seawater.

When plants die, they add carbon to the soil as leaf litter and other dead matter.

THE COST OF DEFORESTATION

Deforestation accounts for about one fifth of greenhouse gas emissions caused by human activities – exceeding emissions from global transport. Halting deforestation and restoring forests that have been cleared could provide about one third of the action needed to combat climate change.

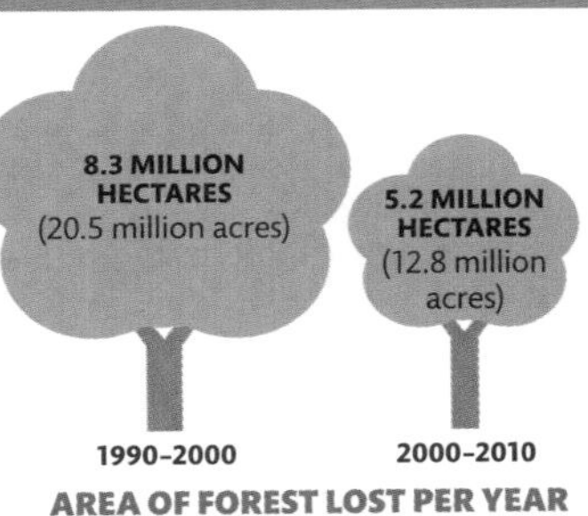

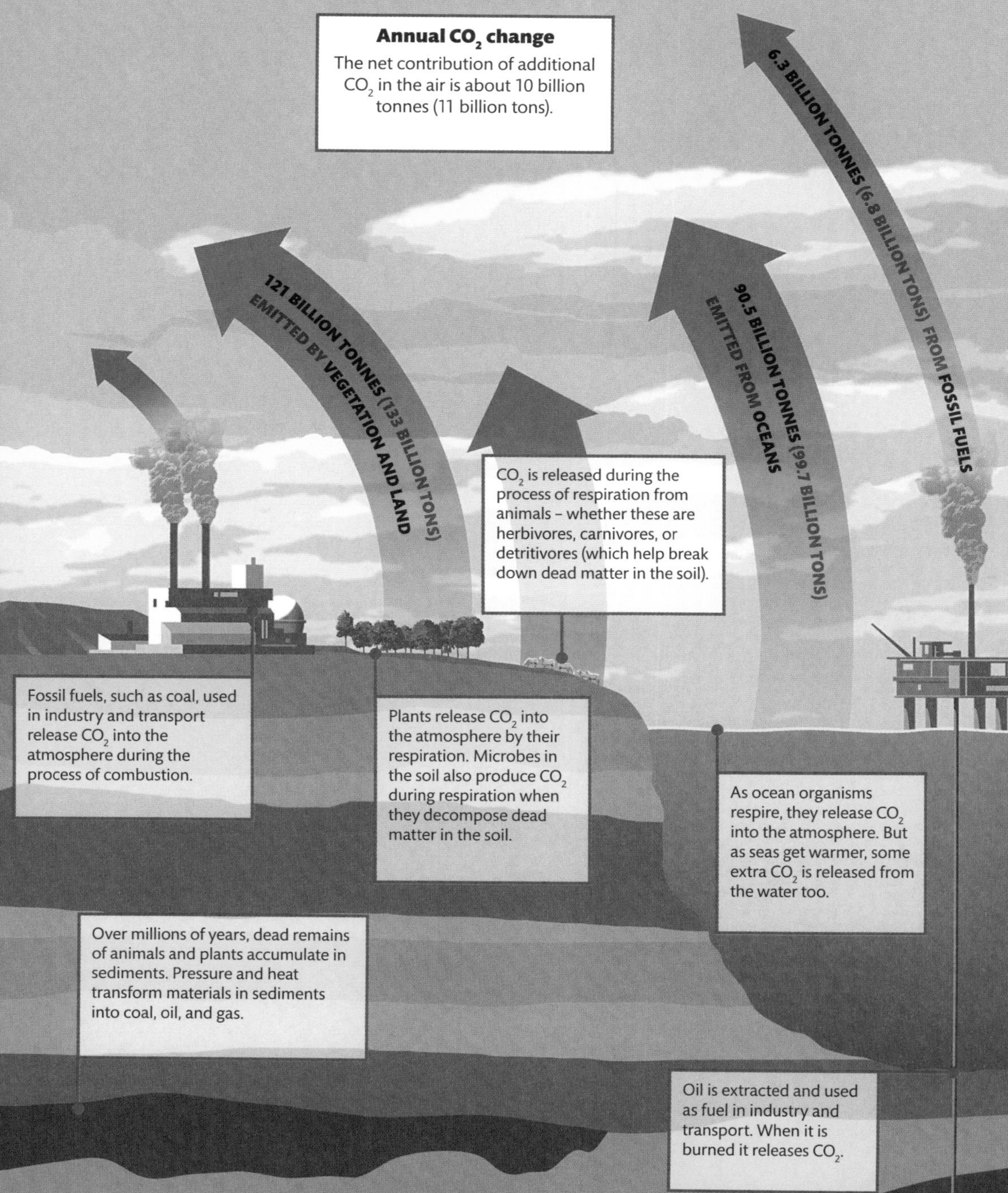
Annual CO_2 change
The net contribution of additional CO_2 in the air is about 10 billion tonnes (11 billion tons).
121 BILLION TONNES (133 BILLION TONS) EMITTED BY VEGETATION AND LAND
90.5 BILLION TONNES (99.7 BILLION TONS) EMITTED FROM OCEANS
6.3 BILLION TONNES (6.8 BILLION TONS) FROM FOSSIL FUELS
CO_2 is released during the process of respiration from animals – whether these are herbivores, carnivores, or detritivores (which help break down dead matter in the soil).
Fossil fuels, such as coal, used in industry and transport release CO_2 into the atmosphere during the process of combustion.
Plants release CO_2 into the atmosphere by their respiration. Microbes in the soil also produce CO_2 during respiration when they decompose dead matter in the soil.
As ocean organisms respire, they release CO_2 into the atmosphere. But as seas get warmer, some extra CO_2 is released from the water too.
Over millions of years, dead remains of animals and plants accumulate in sediments. Pressure and heat transform materials in sediments into coal, oil, and gas.
Oil is extracted and used as fuel in industry and transport. When it is burned it releases CO_2.

Targets for the future

In the 2015 UN Climate Change Conference in Paris, countries confirmed their commitment to limit global warming to less than 2°C (3.6°F) and also agreed to aim for the more challenging limit of 1.5°C (2.7°F).

The UN Framework Convention on Climate Change was adopted at the Earth Summit in Rio de Janeiro, Brazil, in 1992. Negotiations under this legally binding treaty led to a new agreement being forged in Paris in 2015. Under this new agreement, countries adopted voluntary national action plans to reduce greenhouse gas emissions. Although this marked a major step forward, the total cuts are insufficient to meet a 2°C (3.6°F) warming limit. However, a five-yearly review process will require countries to re-examine their ongoing efforts and to consider whether deeper cuts are necessary.

SEE ALSO...

› **The 1.5-degree limit** pp124–25

› **Sustainable development goals** pp184–85

Main polluters

The top 10 biggest emitters of carbon dioxide in 2011 accounted for about two-thirds of global emissions. All these countries (and 175 others) committed to reducing emissions as part of the 2015 Paris climate agreement. In the diagram, the figures are millions of tonnes of CO_2 ($MtCO_2$) emitted in 2011. Proposed cuts are for 2020–2030.

10 Mexico
Plans to cut emissions by 22 per cent by 2030 and will go further if conditions are met, such as a global agreement addressing international carbon price

8 Japan
Despite economic difficulties and nuclear power problems, Japan still aims to cut emissions by 26 per cent compared with 2013

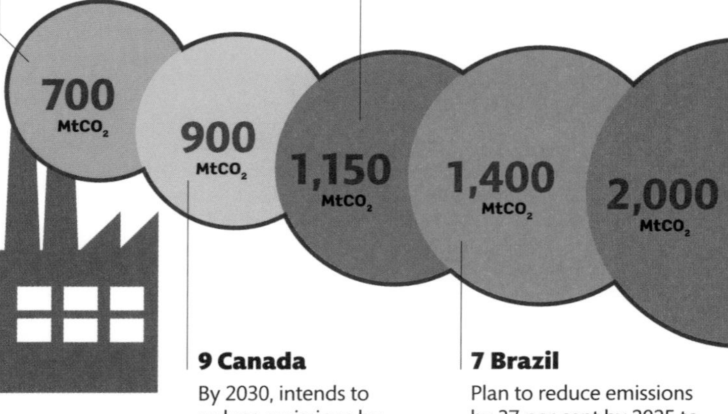

9 Canada
By 2030, intends to reduce emissions by 30 per cent compared with 2005

7 Brazil
Plan to reduce emissions by 37 per cent by 2025 to be met through renewable energy expansion and saving forests

Timeline of change

Since 1992, there have been many significant summits at which countries have debated how best to deal with the challenge of climate change. However, success has been elusive.

1979
First World Climate Conference in Geneva, Switzerland

1988
Intergovernmental Panel on Climate Change (IPCC) founded

1992
UN Framework Convention on Climate Change (UNFCCC) agreed at Earth Summit

1997
Kyoto Protocol, extending the UNFCCC, is signed

2007
China announces its first National Climate Change programme in response to overtaking the USA as the world's largest polluter

"If we act, **we can save the planet** – and we can create millions of jobs ... to **raise the standard of living** for everyone."

JOE BIDEN, 46TH US PRESIDENT

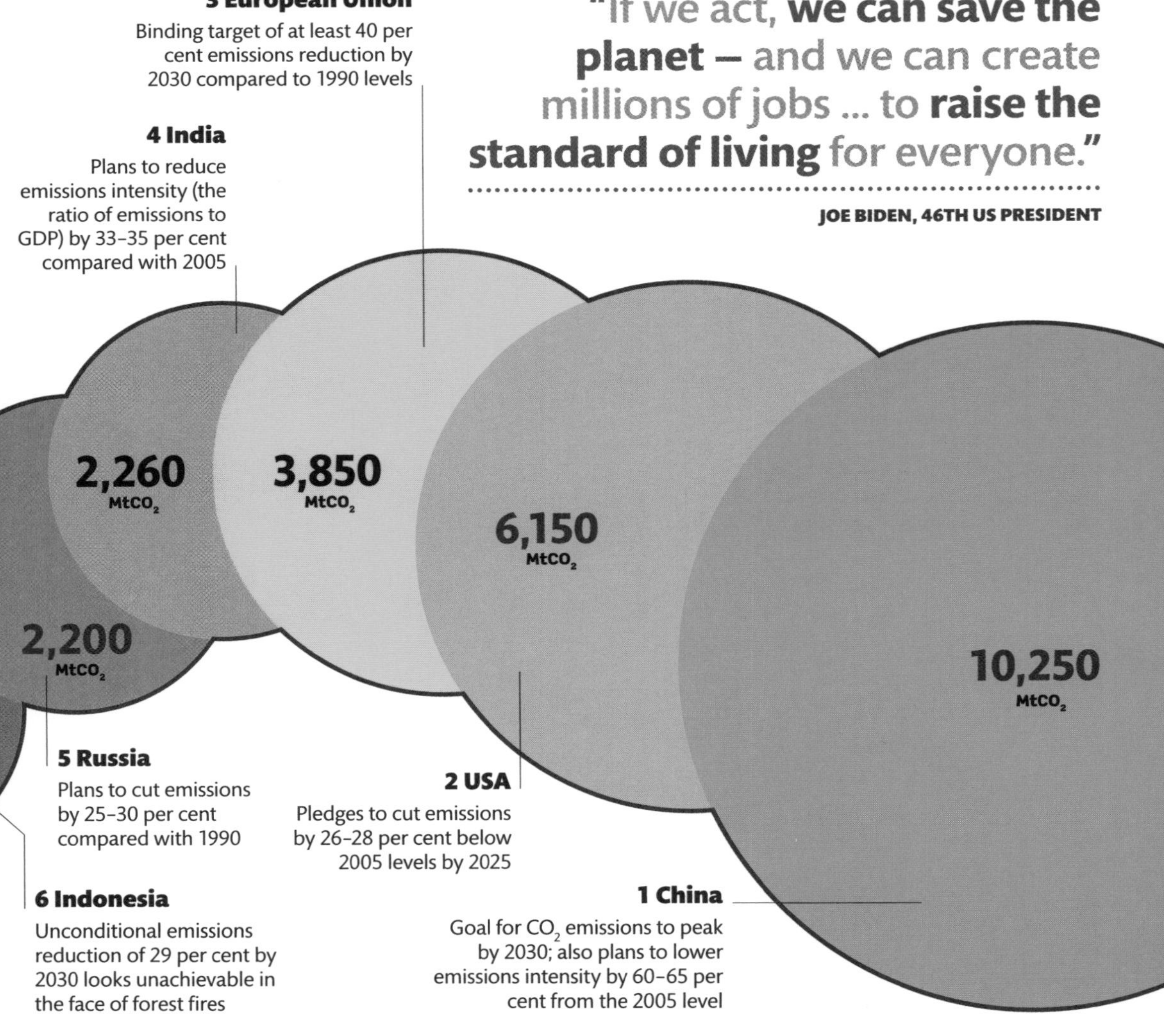
3 European Union
Binding target of at least 40 per cent emissions reduction by 2030 compared to 1990 levels
4 India
Plans to reduce emissions intensity (the ratio of emissions to GDP) by 33–35 per cent compared with 2005
2,260 $MtCO_2$
3,850 $MtCO_2$
6,150 $MtCO_2$
2,200 $MtCO_2$
10,250 $MtCO_2$
5 Russia
Plans to cut emissions by 25–30 per cent compared with 1990
2 USA
Pledges to cut emissions by 26–28 per cent below 2005 levels by 2025
6 Indonesia
Unconditional emissions reduction of 29 per cent by 2030 looks unachievable in the face of forest fires
1 China
Goal for CO_2 emissions to peak by 2030; also plans to lower emissions intensity by 60–65 per cent from the 2005 level

2009 Copenhagen Summit results in a weak, non-binding agreement

2011 Durban climate change talks agree to open negotiations for a new legally-binding climate change treaty to be agreed in Paris in 2015

2014 IPCC Fifth Assessment Report concludes that "human influence on the climate system is clear" and that "human emissions of greenhouse gases are the highest in history"

2015 Paris climate talks agree a global legally-binding commitment to limit the temperature rise to 2°C (3.6°F) and, if possible, 1.5°C (2.7°F)

Toxic air

Air pollution is a major cause of premature death. The rise of huge cities, combined with increased demand for energy and cars, is making things worse.

A wide range of pollutants get into the air and cause damage to human health. Vehicle exhausts, emissions caused by power stations, and forest fires are the principal sources. Common health-threatening pollutants include microscopic particles, oxides of nitrogen, carbon monoxide, and ozone, which is toxic when it is in the air we breathe. Cars and trucks are especially problematic. Nitrogen oxides and particles released from diesel engines, and photochemical smog arising from sunlight acting on petrol exhausts, kill millions.

DAMAGING PARTICLES

Polluting particles are divided into two groups: PM2.5 and PM10, based on their diameter. The WHO sets the maximum safe limit over 24 hours as 25 of the PM2.5 particles per cubic metre of air.

Deaths by disease

Air pollution increases the instances of major diseases. Particles released by combustion, for example, can be less than 2.5 microns in diameter. This means they are small enough to reach the deepest parts of the lungs and cross into the bloodstream. The World Health Organization (WHO) released figures breaking down the 3.7 million pollution-related deaths in 2012 by types of disease.

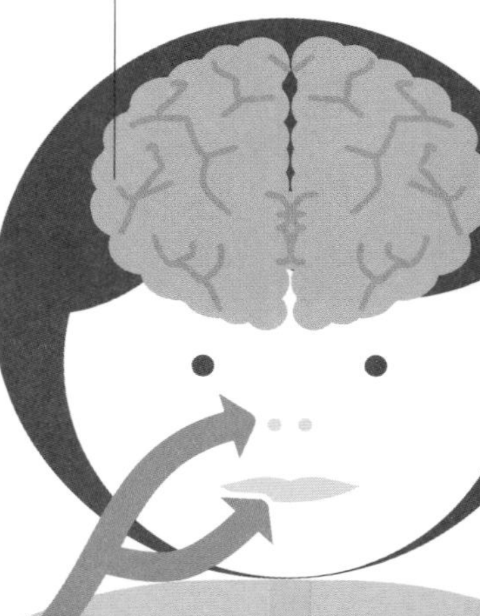

STROKES 40%
Pollutants can cause damage to blood vessels in the brain, causing oxygen starvation in brain tissues and death

Thickness of **human hair** (50–70 microns)

Diameter of **particle PM10** (10 microns), such as dust and pollen

Diameter of the **toxic particle PM2.5** (2.5 microns)

TOXIC PARTICLES

COPD 11%
Chronic obstructive pulmonary disorder (COPD) narrows the airways and can be fatal

HEART DISEASE 40%
Pollution can cause blood vessel damage, restricting the blood flow and triggering heart attacks

LUNG CANCER 6%
The risk rises with increasing exposure to air pollution, including particulate matter

ACUTE LOWER RESPIRATORY DISEASES 3%
The biggest cause of deaths among young children worldwide

Sources of pollution
The main sources of air pollution include power stations, factories, and vehicles. These pollutants are all recognized, but not enough has been done to reduce emissions and millions of people continue to lose their lives.

Most toxic parts of the world

About 88 per cent of deaths caused by air pollution occur in low- and middle-income countries, which between them are home to 82 per cent of the world's population. As of 2012, the Western Pacific and Southeast Asian regions were the worst offenders, with 1.67 million and 936,000 deaths, respectively. Some experts believe that the increasing number of fossil fuel-powered megacities – cities with more than 10 million inhabitants (see pp36–37) will double the number of air-pollution deaths by 2050 compared with 2012. Air quality has actually improved in some parts of the world – the blue regions on the map below signify reductions in air pollution deaths since the 1850s.

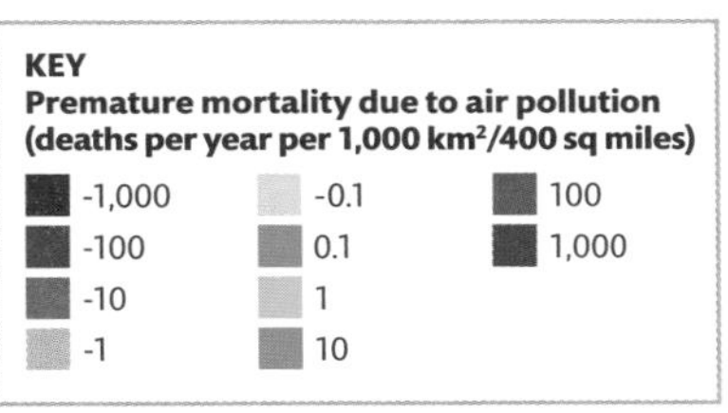

What can we do?

- **Go electric** Choosing an electric vehicle over a petrol or diesel-powered car will do your bit to improve air quality and improve public health.
- **Plant trees** Increasing the number of trees in polluted urban areas can help to clean up the air. Leaves catch particles and other pollutants that are washed to the ground when it rains.

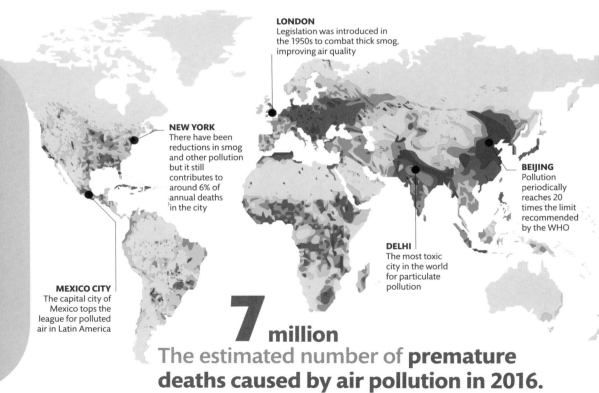

LONDON
Legislation was introduced in the 1950s to combat thick smog, improving air quality

NEW YORK
There have been reductions in smog and other pollution but it still contributes to around 6% of annual deaths in the city

BEIJING
Pollution periodically reaches 20 times the limit recommended by the WHO

DELHI
The most toxic city in the world for particulate pollution

MEXICO CITY
The capital city of Mexico tops the league for polluted air in Latin America

7 million
The estimated number of **premature deaths caused by air pollution in 2016.** 90% were in developing countries.

Acid rain

Acid rain is caused by emissions of sulphur dioxide and nitrogen oxide, which react with water in the atmosphere to produce acids that cause harm to plants, aquatic animals, and buildings. It can also pose serious respiratory problems to humans. The main source of acid rain (and acid snow and sleet) is the large-scale combustion of coal in power stations and industrial plants, such as steel and cement works. Acid rain can travel for hundreds or even thousands of miles. Action taken in some parts of the world, especially in North America and Europe, has reduced the pollutants causing acid rain. It remains a major problem in other countries, however, including China and Russia.

1 Coal is burned in industrial plants and power stations.

2 Acidic particles and gases that don't mix with water droplets in clouds fall to the ground as dry acid precipitation.

Acid rain flows into freshwater systems, polluting lakes and rivers. This makes the rivers and lakes acidic, causing fish and other freshwater life to die.

2 The wind carries pollutants – including sulphur dioxide and nitrogen dioxide – far from their original sources.
4 Acid snow and sleet falls over upland areas. When it melts it flows into rivers and lakes.
3 Sulphur dioxide and nitrogen oxides combine with water in clouds to form weak nitric and sulphuric acid.
4 Acid rain falls.
Trees and plants are damaged, becoming malnourished and prone to disease.
Acid rain corrodes both natural and man-made structures, eating away at paint and buildings.
Acid rain seeps into the ground and dissolves nutrients that are vital to plant health. As the soil becomes more acidic, aluminium is released, which can pollute freshwater.

Changing the land

During the 20th century, the expansion of croplands and of pasture to feed animals, and the development of forestry to sustain a rising demand for timber and paper, has put increasing pressure on the planet. At the same time we have destroyed various ecosystems by chopping down forests and by using land for our own needs at the expense of wildlife. One of the results of this is the desertification of once productive land. Land has become a scarce resource in some countries, and many regions have started to invest in lands far away to produce food and biofuels.

Consuming Earth's natural resources

Scientists have developed an indicator to measure the overall use of Earth's resources called Human Appropriation of Net Primary Production (HANPP). This shows how humans now consume a hugely disproportionate percentage of primary production. (Primary production means the sum of plant biomass produced by photosynthesis.) We use the productive capacity of land by harvesting plant biomass for food or burning it for fuel. This change in land use is the main cause of ecosystem damage and the decline in wildlife diversity and abundance. The main graph shows how our consumption of primary production (HANPP) has increased dramatically over the last century, leaving less to sustain all other species.

"Forests [...] act as **giant global utilities,** providing **essential public services** to the **whole of humanity."**

HRH THE PRINCE OF WALES

Agricultural productivity goes up during the post-war years and less new land is needed to increase food production.

1910 1920 1930 1940 1950

YEAR

BIOMASS CHANGE

One of the most striking indicators of the scale of human's influence on the planet is the huge shift from a world dominated by wild animals to one where humans and farm animals make up the most of the planet's air-breathing vertebrate biomass. Ten thousand years ago 99.9 per cent of vertebrate biomass (total weight) comprised wild animals. With the emergence of agriculture and the domestication of animals this changed. Today, 96 per cent of Earth's vertebrate biomass is comprised of people and farm animals.

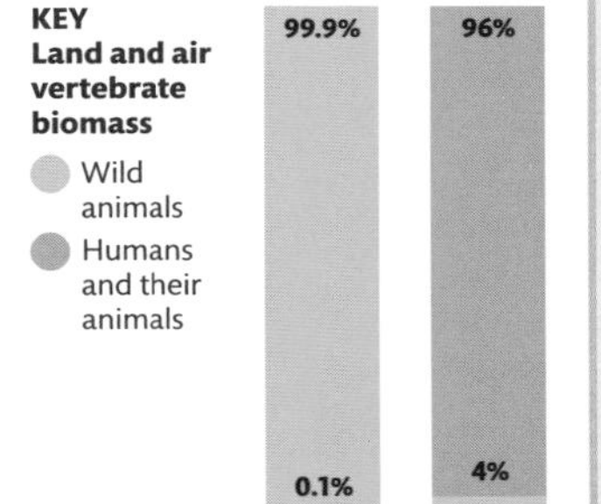

Because of an average increase in crop yields the rate of growth in HANPP levels off, even though population and consumption levels are still increasing.

Rapid population growth is accompanied by a steep increase in human appropriation of land and plant biomass.

1990s
Rapid economic growth in emerging economies leads to more demand for meat and dairy products, and more land to produce them

1960s
Despite more productive farming, a population explosion results in more and more land being harnessed to meet human demand

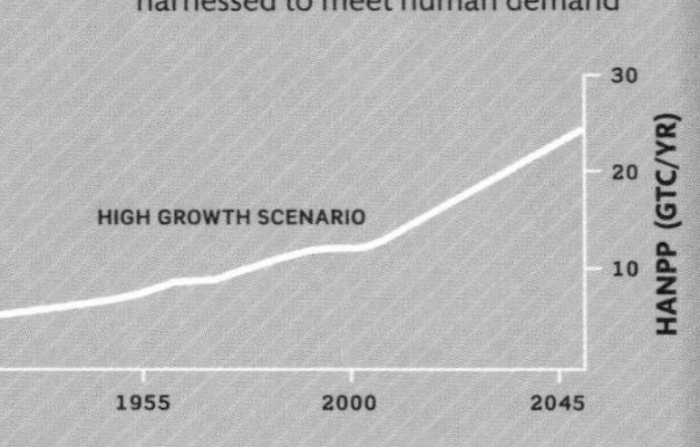

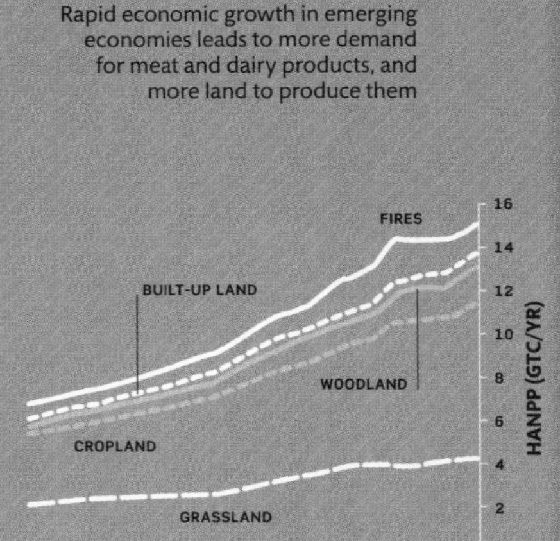

Future trend

Projections based on high growth in bioenergy (such as crops for fuel) suggest that HANPP will increase further up to 2050, creating additional pressures on natural habitats and vital ecosystem services.

Causes for growth

Most growth in HANPP during the last century is explained by natural habitats being converted to cropland and grazing land. Forest fires also account for a substantial proportion, as does the consumption of forest products.

1970 1980 1990 2000

Forest clearance

Most of the world's natural land vegetation has been removed or heavily modified as a result of human activity. The overall global situation is reflected in the drastic reduction in natural forest cover.

Forests are vital to the health of the planet. They are important in capturing greenhouse gases, and for many human needs (see panel, opposite). But, since the beginning of settled agriculture, vast swathes of forest have been lost. Since 1700, the rate of loss has been faster than at any other time in our history. Beginning in Europe and Asia, the process spread to North America and to the tropics. In much of Europe, West Africa, Southeast Asia, and southeastern Brazil, the clearance of natural forests is nearly complete. Agriculture is the main cause of forest loss, often preceded by logging.

Forest loss over time

Until the early 20th century, the highest rates of deforestation were in temperate forests in Asia, Europe, and North America. By the mid-20th century deforestation almost came to a halt in temperate forests but rapidly increased in the tropics. The rate of tropical deforestation remains high. In 2019 it was found that the speed at which forests were being cleared had actually gone up by more than 40 per cent compared with 2014.

KEY
Forest loss
(Millions of hectares/acres)
Temperate forest
Tropical forest

PRE-1700
400 million Ha (1,000 million acres)
10 million Ha (25 million acres)

1700–1849
170 million Ha (420 million acres)
110 million Ha (270 million acres)

1850–1919
140 million Ha (346 million acres)
70 million Ha (173 million acres)

1920–1949
240 million Ha (593 million acres)
100 million Ha (247 million acres)

Winners and losers

In some countries deforestation is occurring rapidly, but in others tree cover is expanding under plantations. These are some of the countries to have recently seen the biggest changes in tree cover.

BIGGEST GAINS
China
Vietnam
Philippines
India
Uruguay

BIGGEST LOSSES
Brazil
Malaysia
Paraguay
Indonesia
Guatemala

WHY PEOPLE NEED FORESTS

Forests are plundered for wood and cleared to make way for farming. While societies gain value from these activities, other and even more important forest values are being lost.

Fuel

Millions of people depend on forests for wood fuel.

Carbon storage

Forests play vital roles in the carbon cycle (see pp132–33), helping to combat climate change.

Water supply

Forests create rain clouds and are vital for water security.

Paper

Forests supply the world with paper.

Soil protection

Woodlands help to limit soil erosion and the spread of deserts.

Flood reduction

Wooded landscapes hold water and help reduce flood risk.

Medicines and food

Many human diseases are treated with drugs first found in forest plants and animals. Forests also provide food.

Biodiversity

About 70 per cent of wildlife diversity on land is found in forests. especially in the tropics.

1950–1979

1980–1995

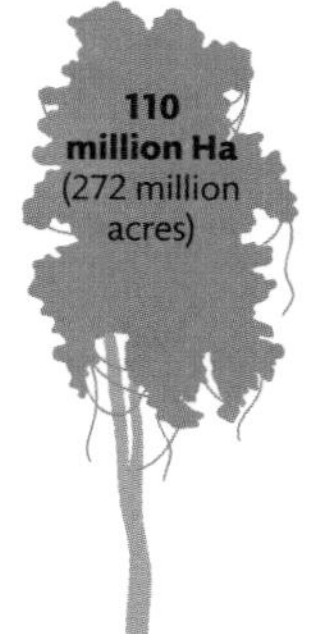

1996–2010

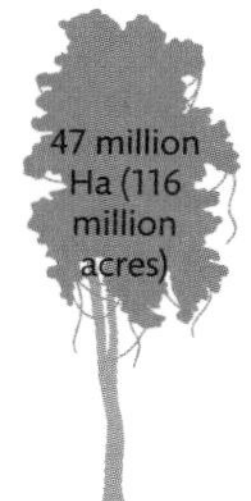

2010–2020

Desertification

Across many of the world's semi-arid regions, the land is turning to desert. This is mostly caused by the degradation of delicate ecosystems, especially savanna woodlands, leading to soil loss and desertification.

Desertification is the persistent degradation of semi-arid dryland ecosystems, such as grasslands and woodlands. It is caused by variations in climate and human activities. More than one third of the world's land area is vulnerable to desertification, with 10 to 20 per cent of all drylands already lost to advancing deserts. The most widespread effects of desertification are seen around the subtropical deserts of North Africa, the Middle East, Australia, southwest China, and western South America. Other areas at risk are the countries surrounding the Mediterranean and Asia's subtropical steppes.

Desertification can cause once productive land to become useless. It is a global issue that has serious implications for biodiversity, poverty eradication, socio-economic stability, and sustainable development.

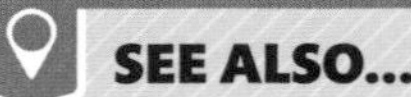

SEE ALSO...

- **Threats to food security** pp68–69
- **Extreme world** pp122–23

CASE STUDY

Lake Chad

- In 1963, Lake Chad in Africa was a vast body of water covering 26,000 km² (10,000 sq miles). In 2001, it was a fifth that size and has since shrunk to 1,300 km² (500 sq miles). Millions of people once relied on the lake for fishing and farming.
- Deforestation, overgrazing, and diverting water for irrigation caused desertification to take hold, impoverishing the people living there.

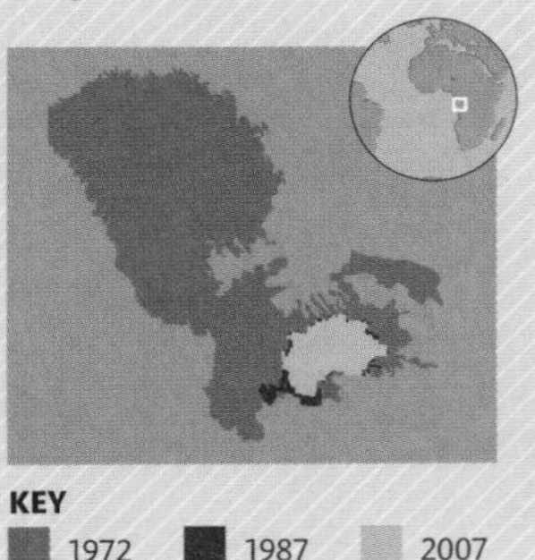

Impacts of desertification

Various human activities, such as deforestation and farming practices, can cause deserts to spread and, in the process, bringing a series of problems. The consequences are being felt in some of the world's most fragile countries, but also more widely. The effects of climate change are aggravating the situation, with droughts exacerbating the more direct human impacts on the land.

Causes

Growing of cash crops

Growing crops for export rather than local markets leads to more intensive farming, causing soil damage.

Incorrect irrigation

Attempts to boost food production with irrigation can cause salt to rise to the top of the soil, making it harder for plants to grow.

Cutting down trees

Felling trees for fuelwood reduces tree cover, leaving soils vulnerable to erosion.

Overgrazing

Too many animals grazing one area for too long remove the vegetation that protects the soil, leading to erosion.

Dry rivers

Damaged soils hold less water and river flow declines. Fewer plants leads to less moisture evaporation to the air and that means less rainfall.

Soil damage

› **Sun-baked, cracked soil** Exposed to the punishing heat of the Sun, soils become baked and impermeable to the scarce rainfall.

› **Soil erosion** With tree cover removed, the soil becomes dry and vulnerable to erosion by wind and water.

Loss of plants and animals

As the desert advances, the dry woodland's native wildlife retreats.

Extreme weather

› **Flash floods** Instead of penetrating into the ground, rainwater runs off the hardened soil crust to cause flash flooding.

› **Gullying** The land is further damaged as flood water concentrates into streams, stripping away the soil to form deep gullies.

› **More sand storms** Loosened soil turns to dust. Windy conditions whip it into the air to form blinding sand storms.

Impact on people

› **Crops and cattle die** As farm animals and crops die, people are made even poorer.

› **Migrants move to cities** As farming is rendered impossible by the march of deserts, people are forced to move to cities.

› **Unrest** Increased demand for services in urban areas causes social tensions.

› **Death** Reduced food production leads to more widespread malnutrition and people die.

Desertification

Physical impacts

Impacts on humans

What can we do?

› **Governments can act** by funding programmes to achieve the aims of the UN Convention to Combat Desertification (agreed in 1992) to improve the living conditions for people in drylands, and maintain and restore the land and soil productivity.

Land rush

Some countries have growing populations but limited potential to grow their own food. Concern about food security has led some governments and investors to seek control of land in other countries.

Lack of suitable land to grow plants for food and biofuels, along with water scarcity, are major issues in a growing number of countries. In the past, trade was used as a means to feed countries with limited land, but now direct control of production is seen as more desirable. In some cases, governments have allocated land to foreign interests without consulting local people, leading to disputes and sometimes violence. As well as creating additional pressure on forests and other natural habitats, the large-scale allocation of land to external agricultural interests also undermines food security in host countries. Two-thirds of large-scale land acquisitions have been in countries with a serious hunger problem.

Land acquisition

The land rush has been a worldwide phenomenon with finance originating from interests in Europe, Middle Eastern nations, Korea, and China to take control of land in Asia, Latin America, and Eastern Europe. It is Africa, however, to which the vast majority of money has been attracted.

KEY
Regions from which investment originated

- Africa
- Asia
- Latin America
- Europe
- North America
- Oceania
- Middle East

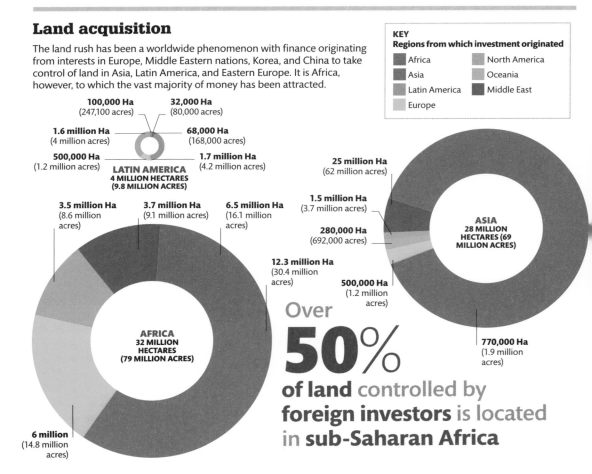

100,000 Ha
(247,100 acres)
32,000 Ha
(80,000 acres)
1.6 million Ha
(4 million acres)
68,000 Ha
(168,000 acres)
500,000 Ha
(1.2 million acres)
1.7 million Ha
(4.2 million acres)

LATIN AMERICA
4 MILLION HECTARES
(9.8 MILLION ACRES)

3.5 million Ha
(8.6 million acres)
3.7 million Ha
(9.1 million acres)
6.5 million Ha
(16.1 million acres)
12.3 million Ha
(30.4 million acres)
6 million
(14.8 million acres)

AFRICA
32 MILLION
HECTARES
(79 MILLION ACRES)

25 million Ha
(62 million acres)
1.5 million Ha
(3.7 million acres)
280,000 Ha
(692,000 acres)
500,000 Ha
(1.2 million acres)
770,000 Ha
(1.9 million acres)

ASIA
28 MILLION
HECTARES (69
MILLION ACRES)

Over
50%
of land controlled by **foreign investors** is located in **sub-Saharan Africa**

What can we do?

- **Governments and investors** must prioritise the interests of local populations before making decisions to change land ownership or control.
- **Investors** must ensure that their activities contribute to the sustainable development and food security in host countries.
- **Involve people** living in affected areas in discussions that could lead to a change in ownership of the land they depend on.

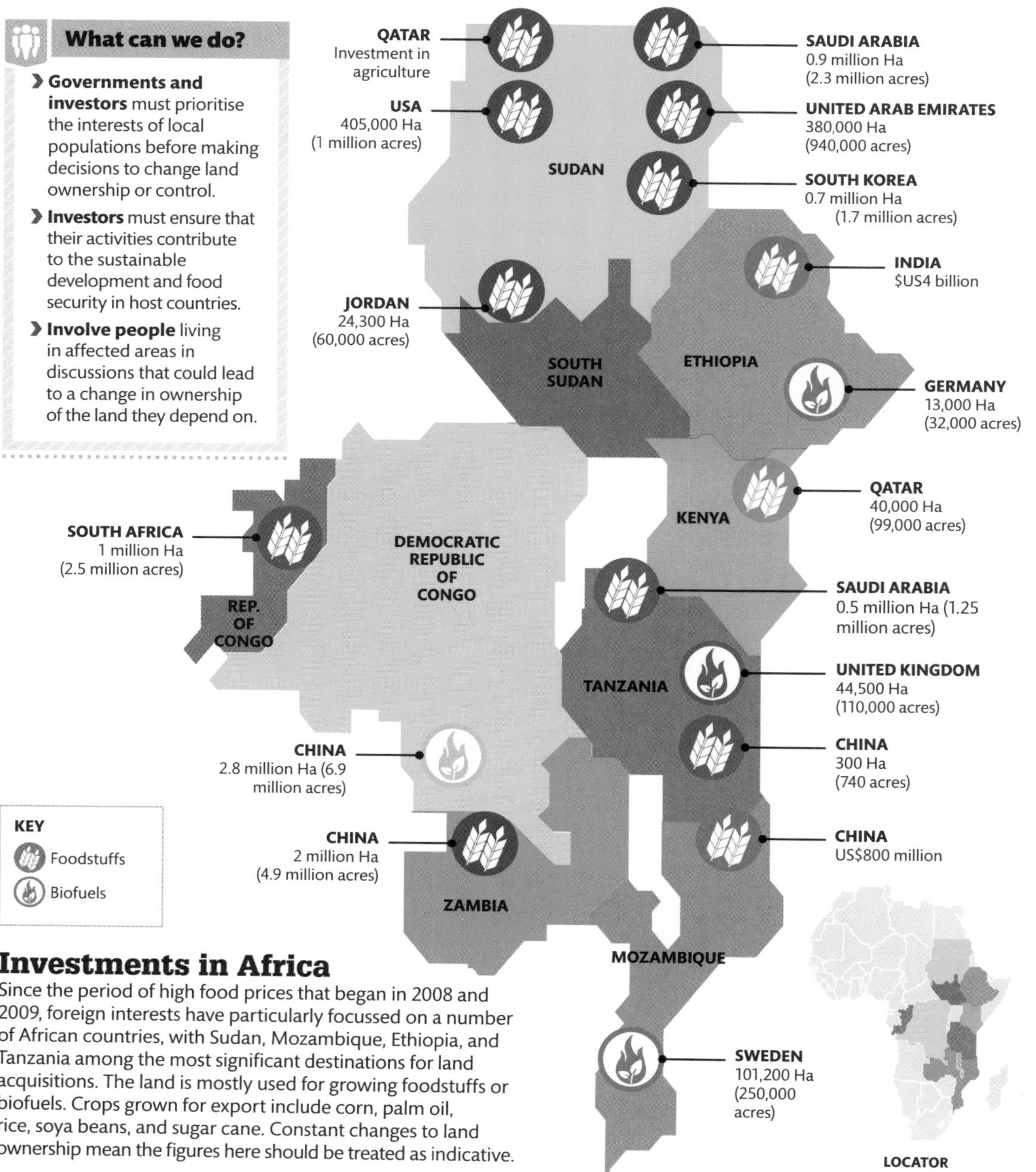

Investments in Africa

Since the period of high food prices that began in 2008 and 2009, foreign interests have particularly focussed on a number of African countries, with Sudan, Mozambique, Ethiopia, and Tanzania among the most significant destinations for land acquisitions. The land is mostly used for growing foodstuffs or biofuels. Crops grown for export include corn, palm oil, rice, soya beans, and sugar cane. Constant changes to land ownership mean the figures here should be treated as indicative.

Sea changes

Fish caught from the seas are a vital source of economic development. Global catches of fish contribute an estimated US$278 billion per year to the global economy, and US$160 billion more comes from boat-building and other related industries. Global wild fish stocks provide employment for hundreds of millions of people, the vast majority of whom live in developing countries. The fishing industry contributes to global food security – about one billion people are reliant on wild-caught fish for their main source of protein. Sustaining these benefits depends on sustaining fish stocks.

Plundering the oceans

During the 1950s, marine fish catches grew rapidly. This was due to bigger vessels fishing in greater numbers, as well as the use of new technologies, including sonar equipment. Government subsidies gave incentives for over-fishing, so today more than half of stocks are at their maximum sustainable yield – the largest catch that can be taken – and about a third of them are overexploited, some to the point of collapse. This graph charts annual global fish landings from marine waters from 1950 until 2018. The World Bank estimates that if fish stocks were better managed they could generate US$50 billion more economic value each year.

"If you're **overfishing** at the top of the food chain, and **acidifying the ocean** at the bottom, you're creating a squeeze that could conceivably **collapse the whole system.**"

TED DANSON, AMERICAN ACTOR AND OCEAN CAMPAIGNER

1950 1955 1960 1965 1970 1975 1980

YEAR

FISH UNDER THREAT

Many organizations, including the UK's Marine Conservation Society and the USA's Environmental Defence Fund, offer advice on which fish to eat. They discourage consumption of threatened species, such as bluefin tuna and sturgeon, and encourage people to choose herrings, mackerel, and other species from healthy stocks. The Marine Stewardship Council certifies sustainable fish to help consumers make good choices.

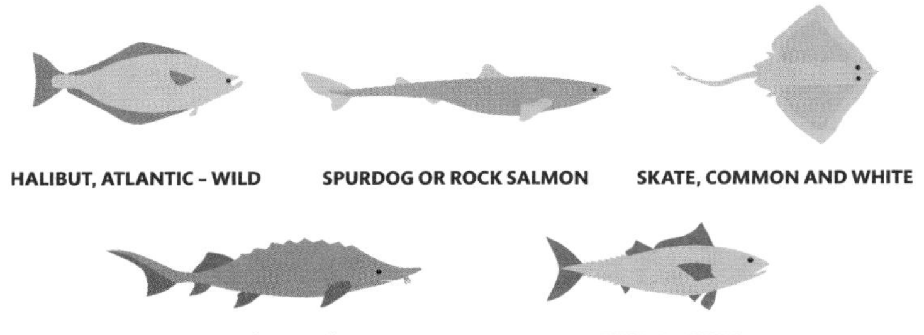

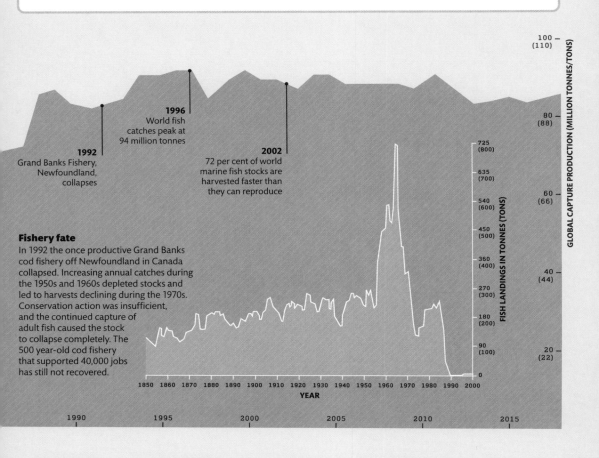

1992
Grand Banks Fishery, Newfoundland, collapses

1996
World fish catches peak at 94 million tonnes

2002
72 per cent of world marine fish stocks are harvested faster than they can reproduce

GLOBAL CAPTURE PRODUCTION (MILLION TONNES/TONS)

100 (110), 80 (88), 60 (66), 40 (44), 20 (22)

1990, 1995, 2000, 2005, 2010, 2015

Fishery fate

In 1992 the once productive Grand Banks cod fishery off Newfoundland in Canada collapsed. Increasing annual catches during the 1950s and 1960s depleted stocks and led to harvests declining during the 1970s. Conservation action was insufficient, and the continued capture of adult fish caused the stock to collapse completely. The 500 year-old cod fishery that supported 40,000 jobs has still not recovered.

FISH LANDINGS IN TONNES (TONS)

725 (800), 635 (700), 540 (600), 450 (500), 360 (400), 270 (300), 180 (200), 90 (100), 0

1850, 1860, 1870, 1880, 1890, 1900, 1910, 1920, 1930, 1940, 1950, 1960, 1970, 1980, 1990, 2000

YEAR

Farming fish

As pressure on wild fish stocks has increased, there has been a rapid expansion in farmed-fish production. While this has made a significant contribution towards meeting nutrition and food security goals, it has also created new challenges.

During the past 50 years, the expansion of fish farming, also known as aquaculture, has been dramatic. Whereas in 1970 only 5 per cent of food fish were sourced from farms, today farmed fish make up about half of all fish eaten in the world. That proportion is expected to rise to nearly two-thirds by 2030.

Fish farming today is a global industry, supplying massive quantities of both marine and freshwater fish, including cod, salmon, bass, and catfish. Aquaculture operations also supply increasing amounts of crustaceans, such as prawns and lobsters, and molluscs, like mussels.

Growth in farmed fish production between 1980 and 2010 outpaced growth in the wild fish catch – so much so that the average consumer in 2010 ate almost seven times more farmed fish than in 1980. Fish are relatively efficient at converting feed to protein for human consumption, but a number of environmental issues have accompanied the rise of farmed fish.

60%

China's share of the world's farmed fish production

Aquaculture's impacts

Fish farming has led to a significant increase in the availability of healthy protein. A number of environmental impacts have emerged as production as risen, however, including the spread of parasites to wild fish – even though farmed fish are kept inside nets or cages.

Fish and fish oil

Species such as salmon are fed on smaller fish, including young wild-caught species.

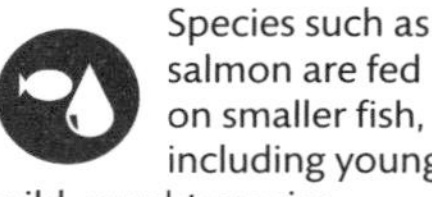

Habitat loss

Creation of fish farms can cause habitat damage. Many areas of ecologically important mangrove forests have been cleared to make way for shrimp farms.

Parasites

Parasites such as lice can quickly spread through confined numbers of captive fish populations, then pass into the surrounding environment to infect wild fish species.

Water quality

Substances added to maintain the health of captive fish, such as antibiotics, flow out and affect marine ecosystems.

Waste pollution

Uneaten food and fish faeces degrade, depleting oxygen, killing plants and animals.

THE RISE OF FARMED FISH

Total farmed fish output was about one million tonnes (1.1 million tons) in 1960. In 2018, 82 million tonnes (90 million tons) of farmed fish was produced: 51 million (56 million tons) from freshwater and 31 million (34 million tons) from the sea. As well as being about equivalent to the volume of wild caught sea fish, fish farming also employed nearly 60 million people.

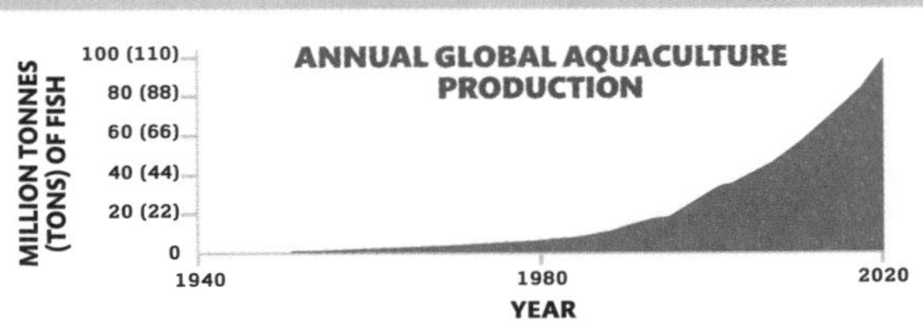

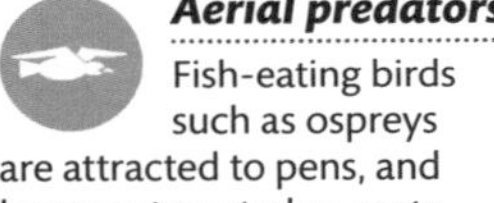

Aerial predators

Fish-eating birds such as ospreys are attracted to pens, and become targeted as pests.

Drugs

Antibiotics are used to prevent and treat diseases. Growth hormones and pigments may be added.

Herbicides

Herbicides are often added to combat algal overgrowth in or near farming pens.

Diseases

Massed numbers of fish in confined spaces create an ideal environment for incubating disease, which can pass to wild fish.

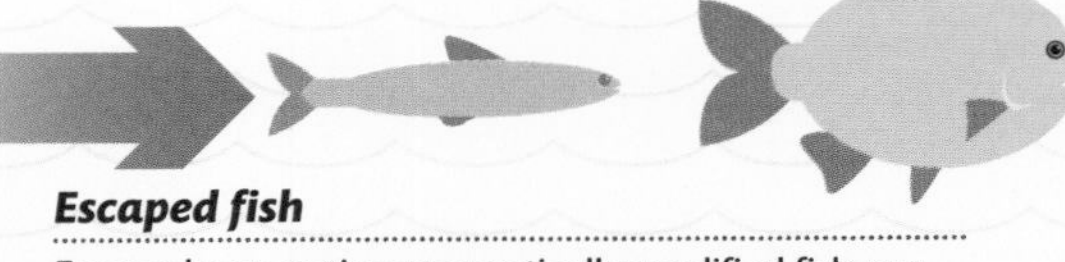

Escaped fish

Escaped non-native or genetically modified fish can cause ecological impacts, competing with wild fish for food, preying on wild fish, passing on diseases, and interbreeding with native populations.

Underwater predators

Fish-eating seals, sharks, and dolphins can become tangled in nets and can be killed during attempts to catch the fish inside.

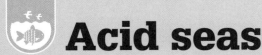

Acid seas

Up to half of the carbon dioxide released because of human activities has been absorbed by the oceans. This has caused marine environments to rapidly become more acidic, leading to conditions not experienced on Earth for more than 20 million years. This has had profound impacts on many ecologically vital species, including oysters, clams, urchins, corals, and plankton. The decline of these and other organisms will cause disruption to entire food webs, bringing devastating consequences for industries dependent on fish and shellfish. Progressive acidification will also limit the oceans' ability to store carbon, as animals that use carbonate to make their shells decline.

Pre-industrial world (1850)

Lower levels of atmospheric carbon dioxide (CO_2) were absorbed by seawater in pre-industrial times. Since then, its acidity has risen by 30 per cent, equivalent to a drop of 0.1 pH unit, caused by emissions from fossil fuels and deforestation.

carbon dioxide

Lower pre-industrial atmospheric CO_2 levels made the ocean's water less acidic, so its pH was higher: about 8.2, compared with 8.1 today.

In less acidic seas (associated with lower CO_2 levels) coral and other animals can easily extract dissolved carbonate from the water to make their exoskeletons and shells.

Healthy oceans maintain good fish stocks.

Future trend (2100)

If CO_2 emissions remain unchecked, by 2100 the acidity of seawater is projected to rise even further: by 150 per cent more than it is today – equivalent to another drop of 0.4 pH unit.

increased levels of carbon dioxide

THE CHEMISTRY OF ACIDIFICATION

When carbon dioxide (CO_2) dissolves in water (H_2O) the two molecules react together to form carbonic acid (H_2CO_3). Carbonic acid then splits to release hydrogen ions (H^+) and hydrogen carbonate ions (top right). The more hydrogen ions in water, the more acidic it is and the lower its pH. Hydrogen ions react with carbonate in the seawater (bottom right), so less carbonate is available for making shells. They also react with carbonate in existing shells, making them corrode.

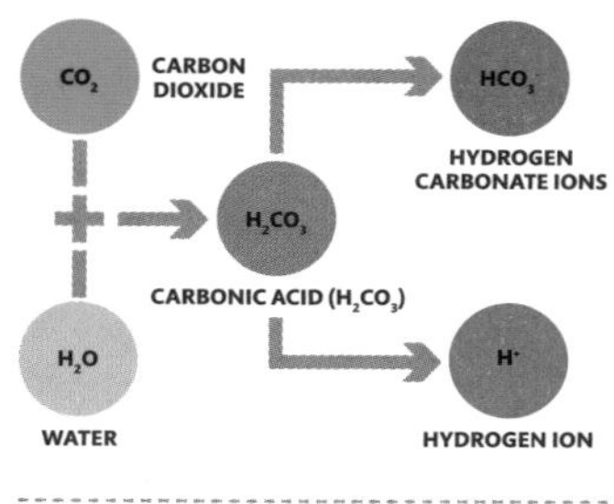

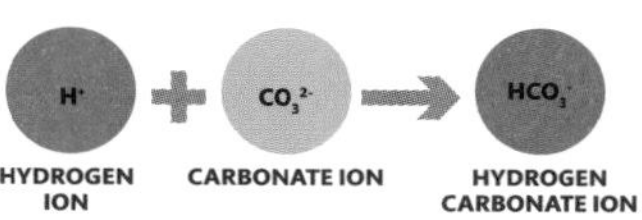

Higher future atmospheric CO_2 levels will make the ocean's water more acidic, so its pH will drop to about 7.7.

Jellyfish are tolerant of warmer and more acid seas. They compete with other sea creatures for food and eat fish eggs. Jellyfish species have spread and numbers have increased dramatically in many areas of oceans.

healthy pteropod shell

acid seas dissolve pteropod shells

Pteropods are small free-swimming sea snails. Lab experiments have shown that their shells take little more than six weeks to corrode in seawater with the same acidity as that projected for 2100.

Coral skeletons become fragile, changing shape and crumbling, and are unable to reproduce. Entire reefs may disintegrate in more acidic seas.

Dead seas

High levels of pollutants in the ocean can have a devastating impact on marine life. Substances such as nitrogen and phosphorus act as fertilizers, triggering a process called eutrophication, which removes oxygen from seawater and creates so-called dead zones.

If nitrogen- and phosphorus-rich agricultural fertilizers, animal waste, detergents, or sewage leak into waterways, contaminated freshwater ends up in the sea, where it can create a dead zone. Dead zones are particularly prevalent in coastal waters, where major rivers discharge, and have such low oxygen levels they no longer sustain life. They cause many damaging effects, from loss of wildlife biodiversity to fisheries collapse. The situation is reversible if the cause is halted and the area is supplied with oxygenated water.

How dead zones form

Eutrophication can occur in any water body, including lakes, rivers, or seas. It usually happens when an excess of nutrients runs into the water from surrounding land controlled by human activity, such as farmland, golf courses, and lawns, all of which are heavily fertilized.

CASE STUDY

The Gulf of Mexico dead zone

- Almost half the continental USA drains into the Mississippi River. As it flows into the Gulf of Mexico, the river creates a vast dead zone each spring, due to seasonal runoff of agricultural fertilizer. In 2015, this oxygen-starved area extended almost 17,000 sq km (6,500 sq miles). Marine life cannot survive in waters where oxygen levels are below 2 mg/litre.

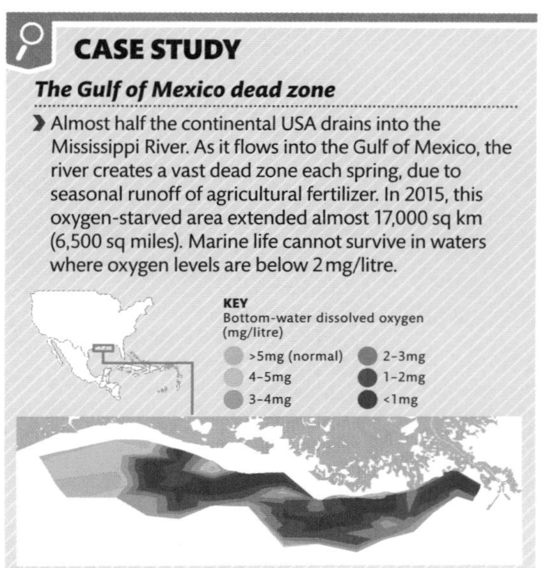

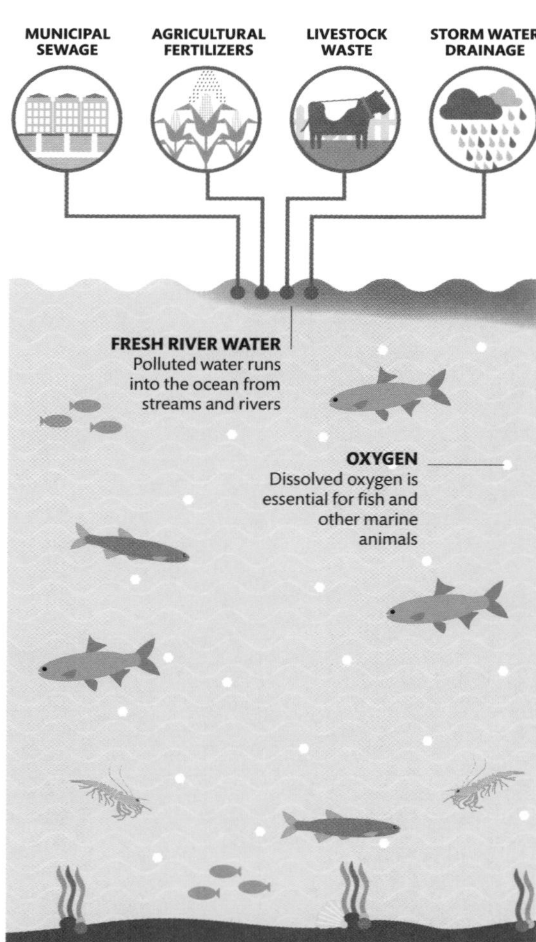

Contaminated water flows in

Water rich in nutrients (from sewage and fertilizers, for example) flows into the sea and forms a layer above denser saltwater.

405

The total number of **dead zones** in coastal waters **worldwide**

What can we do?

- **Prevent untreated wastewater** from being channelled into rivers and seas.
- **Limit the use** of industrial fertilizers in problem areas, such as along coastlines and major rivers.
- **Restore wetlands** and natural coastal defences, which help to filter nutrients out of the water before it reaches the sea.

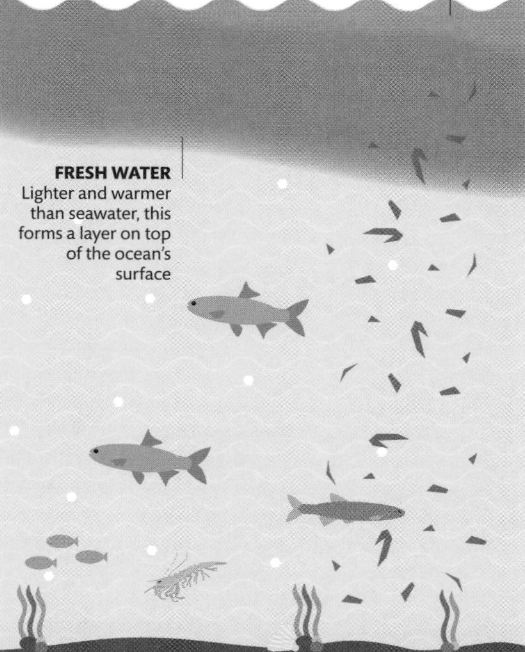
SUNLIGHT WARMS SURFACE
ALGAL BLOOM
Fuelled by sunlight and fertilizers, large areas of algae form, blocking sunlight from water plants
FRESH WATER
Lighter and warmer than seawater, this forms a layer on top of the ocean's surface

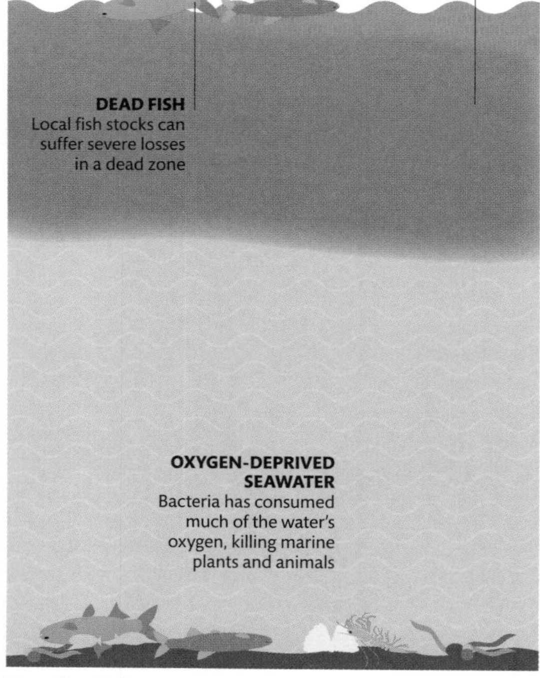
FRESH WATER
As more nutrient-rich fresh water flows into the area, the dead zone expands
DEAD FISH
Local fish stocks can suffer severe losses in a dead zone
OXYGEN-DEPRIVED SEAWATER
Bacteria has consumed much of the water's oxygen, killing marine plants and animals

Algae thrive in fresh water layer

Warm sun provides perfect conditions for algae to form. At the end of their life cycles, dead algae sink to the sea floor where they decompose. In this process, oxygen is removed from the water.

Death of the ecosystem

Low oxygen levels cause marine animals to leave, mutate, or die. Increased decomposition of dead matter exacerbates the lack of oxygen in the water, and the dead zone is formed.

Plastic pollution

Packaging, consumer products, and fishing nets are among the plastic items discarded in the oceans. These kill sea creatures, while plastic particles concentrate pollutants and enter food chains via filter-feeding plankton.

Most plastic now in the oceans was originally dumped on land and entered the marine environment via rivers. About 80 million tonnes (88 million tons) of plastic litter is already in the seas and about eight million more plastic items are added each day. The quantity of plastic debris is rising fast as more people embrace consumer lifestyles. Some wildlife species mistake floating plastic for food, and each year millions of animals and birds die as a result. The United Nations Environment Programme estimates that the impact of plastic pollution on marine life costs the global economy US$13 billion every year.

Deadly gyres

Gyres are large areas of open ocean where slow-moving currents converge. Light plastics are carried on these currents into the gyres, where they are concentrated and held in vast areas of drifting plastic waste. There are five main gyres, including the North Pacific Ocean. A vast quantity of plastic debris drifts in the centre of this gyre. Another is in the Bay of Bengal, where plastics are fed into the sea via Asia's largest rivers, including the Ganges.

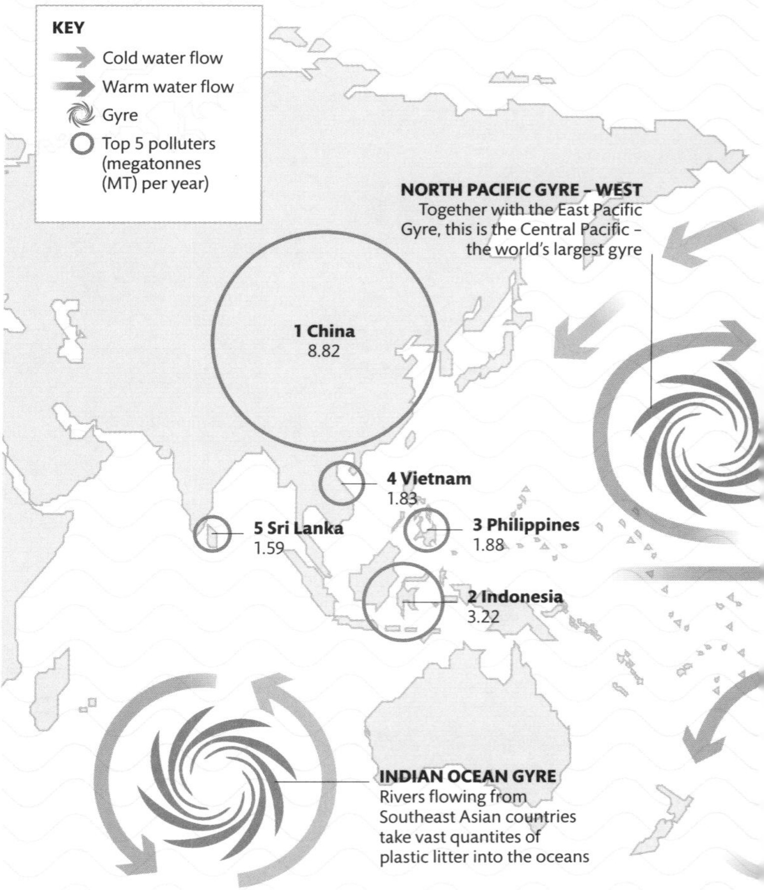

What can we do?

- **Restrict the sale** of single-use plastics, such as supermarket bags.
- **Encourage deposit schemes** for plastic bottles.
- **Invest in** solid waste and recycling facilities.
- **Developing countries** should invest in modern recycling.

What can I do?

- **Stop buying plastic –** choose reusable alternatives.

KEY
Cold water flow
Warm water flow
Gyre
Top 5 polluters (megatonnes (MT) per year)

NORTH PACIFIC GYRE – WEST
Together with the East Pacific Gyre, this is the Central Pacific – the world's largest gyre

1 China 8.82

2 Indonesia 3.22

3 Philippines 1.88

4 Vietnam 1.83

5 Sri Lanka 1.59

INDIAN OCEAN GYRE
Rivers flowing from Southeast Asian countries take vast quantites of plastic litter into the oceans

90% of all litter floating on the ocean's surface is plastic

PLASTIC BREAKDOWN

It can take many years, or even centuries, for plastic debris to break down. Microscopic plastic particles, broken down from larger pieces of litter, attract toxic chemicals that enter the food chain and cause harm.

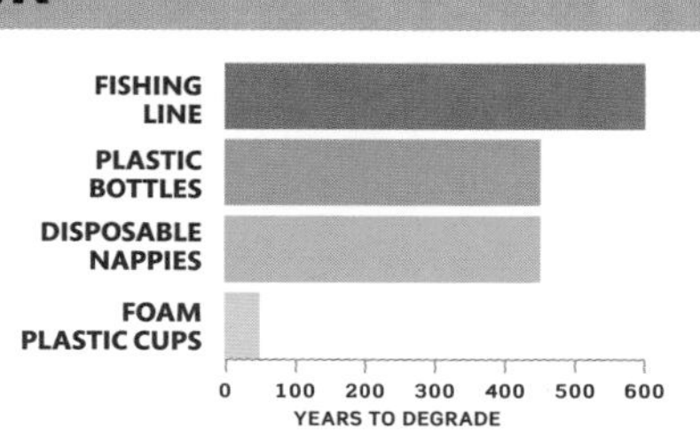

NORTH ATLANTIC GYRE
Stretches from near the equator almost to Iceland, and from the east coast of North America to the west coasts of Europe and Africa

NORTH PACIFIC GYRE – EAST
In parts of this gyre, there are one million pieces of rubbish per square kilometre

SOUTH PACIFIC GYRE
Despite being farthest from any continents and productive ocean regions, the South Pacific Gyre still has a lot of plastic drifting in it

SOUTH ATLANTIC GYRE

EFFECT ON WILDLIFE

Plastic debris has a huge impact on wildlife – either directly or indirectly, as these examples show.

Birds

High mortality among young birds occurs in many albatross colonies because the babies are fed on plastic items, including discarded lighters found drifting in the sea.

Turtles

Some plastic litter, such as fishing nets, lines, and plastic bags, can entangle animals including turtles, dolphins, and birds, causing them to drown.

Plankton

Micro-particles of plastic are taken up both by plankton and by plankton-feeding animals, causing problems for their digestion.

Whales and dolphins

Plastic ingestion has been noted in 56 per cent of whale, dolphin, and porpoise species. Whales have also mistaken plastic bags for squid. One whale was found with 17 kg (37 lb) of plastic in its body.

The great decline

The disappearance of wildlife species is perhaps the most pressing and serious of all environmental problems, threatening the loss of valuable natural "services" (see pp164–65) and, as a result, undermining human wellbeing. A number of stresses are causing the disappearance of natural diversity on a scale not seen for 65 million years, since the extinction of the dinosaurs. The already accelerating rate of species loss is set to become faster still, as existing pressures arising from human population growth, expansion of farming, and economic development become more intense.

Disappearing wildlife

Animal extinctions caused by humans began tens of thousands of years ago, when large mammals, including woolly mammoths and cave lions, were hunted to oblivion by bands of hunter-gatherers. Since then, other pressures have been added to the effects of hunting. During the age of European exploration and colonization, many aggressive invasive species of animals and plants were moved around the world, causing extinctions among native species (see pp162–63). Today, planet-wide degradation of the terrestrial biosphere (see pp140–41) is the main driver of species loss.

> **"We are undoubtedly exterminating species at a speed which has never been known before."**
>
> **SIR DAVID ATTENBOROUGH, BRITISH BROADCASTER AND NATURALIST**

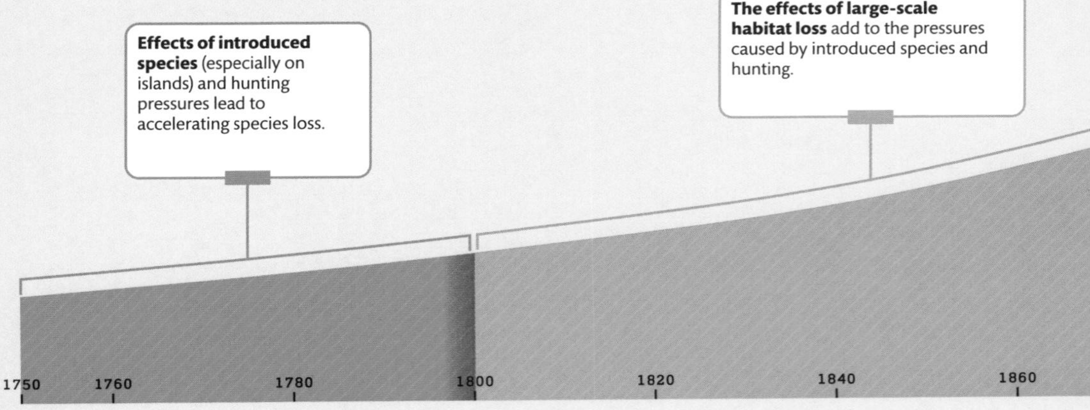

Effects of introduced species (especially on islands) and hunting pressures lead to accelerating species loss.
The effects of large-scale habitat loss add to the pressures caused by introduced species and hunting.
1750
1760
1780
1800
1820
1840
1860
YEAR

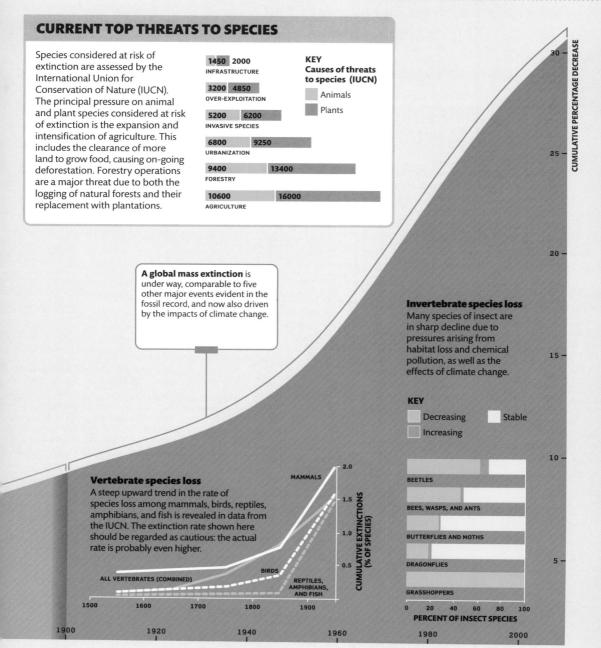

CURRENT TOP THREATS TO SPECIES

Species considered at risk of extinction are assessed by the International Union for Conservation of Nature (IUCN). The principal pressure on animal and plant species considered at risk of extinction is the expansion and intensification of agriculture. This includes the clearance of more land to grow food, causing on-going deforestation. Forestry operations are a major threat due to both the logging of natural forests and their replacement with plantations.

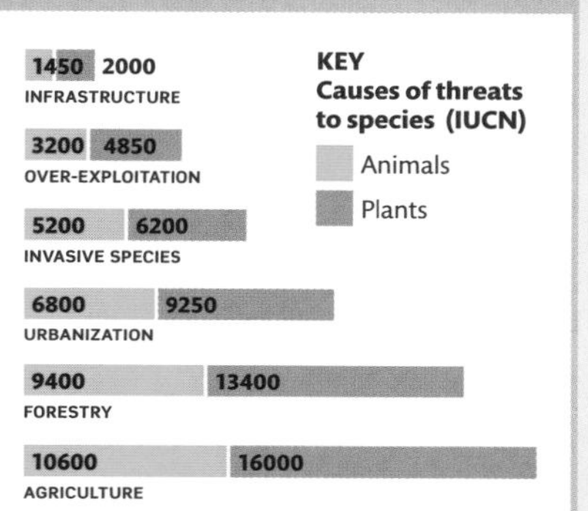

A global mass extinction is under way, comparable to five other major events evident in the fossil record, and now also driven by the impacts of climate change.

Invertebrate species loss
Many species of insect are in sharp decline due to pressures arising from habitat loss and chemical pollution, as well as the effects of climate change.

Vertebrate species loss
A steep upward trend in the rate of species loss among mammals, birds, reptiles, amphibians, and fish is revealed in data from the IUCN. The extinction rate shown here should be regarded as cautious: the actual rate is probably even higher.

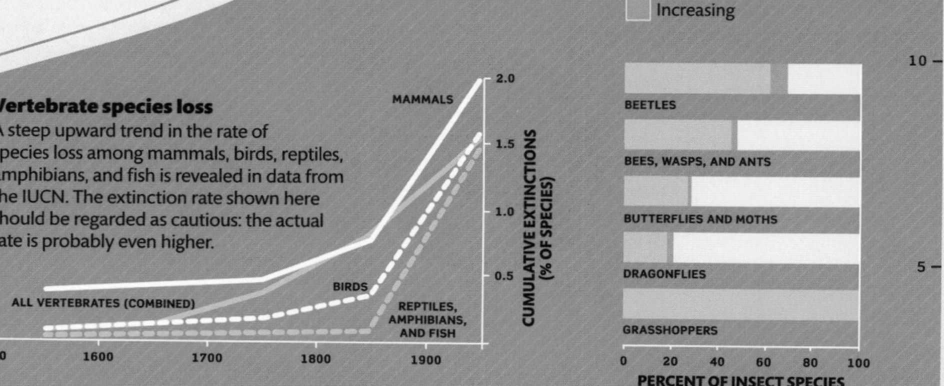

Biodiversity hotspots

The diversity of wildlife species on Earth is not evenly spread. Some places have a far richer diversity of animals and plants. But many such areas are under threat. These areas are known as biodiversity hotspots.

Biodiversity hotspots are places where nature is most diverse and unique, but also where it is most under pressure. Natural diversity sustains human welfare in a multitude of different ways. All of our food and many of our medicines are derived from wild species. There is also huge potential benefit from the process of biomimicry – this is the idea of copying other life forms to find solutions to, for example, engineering and design challenges. By permitting these unique areas to be damaged through deforestation, for example, and allowing species to become extinct we risk losing these benefits that nature provides. Conserving the remaining natural habitats in these biodiversity hotspots is therefore vital not only for conserving wildlife but also for protecting humanity's future prospects.

Caribbean Islands

The islands of the Caribbean form a major hotspot with a range of habitats from 3,000-m (10,000-ft) peaks to low-lying deserts. They are home to 6,550 native plant species and more than 200 threatened endemic vertebrates.

MADREAN PINE-OAK WOODLANDS

CALIFORNIA FLORISTIC PROVINCE

MESOAMERICA

CERRADO

TUMBES-CHOCO-MAGDALENA

TROPICAL ANDES

VALDIVIAN FORESTS

Where nature is most diverse

Conservation International has identified 35 hotspots. Together they cover only 2.3 per cent of the Earth's land surface, and yet more than 50 per cent of the world's plant species and 42 per cent of all terrestrial vertebrates are found in these areas. All of these hotspots are threatened by human activities. As a whole, more than 70 per cent of the natural vegetation has already been lost. Deforestation is a major pressure, caused by the expansion of farming, logging, and mining.

Atlantic Forest

The Atlantic Forest stretches along Brazil's coast. Long isolated from other major rainforest blocks in South America, the Atlantic Forest has an extremely diverse and unique mix of vegetation and forest types, including around 8,000 native plant species. Centuries of logging, cattle ranching, mining, and clearance for sugar cane plantations has devastated this unique habitat.

93% HABITAT LOST

7% LEFT

LOSS OF ATLANTIC FOREST SINCE 1500

More than 70% of natural vegetation has been lost across 35 hotspots

Caucasus

This region includes a range of important habitats, such as grassland, desert, swamp forests, arid woodlands, broadleaf forests, montane coniferous forests, and shrublands. Together they are home to around 1,600 native plant species.

Sundaland

This western half of the Indo-Malayan archipelago has two of the world's largest islands – Borneo and Sumatra. Isolated by sea level rise, the rainforests on these and other islands support many unique species, such as the critically endangered Sumatran tiger. Deforestation threatens the 15,000 native plant species, and subsequent habitat loss threatens 162 endemic vertebrates here.

SUMATRAN TIGER

Cape Floristic Region

On the southwestern tip of the African continent lies an exceptionally diverse region of shrublands, including the flower-rich fynbos. This unique habitat contains 6,210 native plant species.

Southwest Australia

In this region of Australia lies a mixture of Eucalyptus woodlands, thickets, scrub-heath, and heath. This supports some 2,948 plant and 12 threatened vertebrate species that occur nowhere else.

What can we do?

› **Retaining natural habitats** in the hotspots will require the legal protection of at least the best-quality areas, with all rules adopted to protect habitats and wildlife fully enforced. It will also be necessary to find ways for farmers to make a living without encroaching into natural areas.

What can I do?

› **Make regular visits** to areas that are protected for nature, both near to home and when you are travelling. The more that protected areas are used, whether they are diversity hotspots or not, the bigger the incentive for governments and individuals to work to keep them intact.

Invasive species

The spread of species to places where they are not native can cause serious disruption to local ecosystems. The arrival of these invasive alien species can lead to the decline or extinction of native wildlife.

At the global level, the impact of so-called invasive alien species may be as damaging to ecosystems and wildlife diversity as the effects of habitat loss and degradation. Thousands of species have already been driven to extinction by animals and plants moved around by people. Sometimes species are deliberately introduced, such as rabbits to Australia, where the damage they caused to native vegetation led to the decline of many of that continent's birds and mammals.

Other species were taken to new places inadvertently. Many flightless birds once confined to single islands have been driven to extinction because of predation by rats arriving on ships.

SEE ALSO...

› **Biodiversity hotspots** pp160–61

› **Nature's spaces** pp190–91

Invasive species on land

Predation, spread of disease, and competition for food are among factors that lead non-native species to displace native animals and plants. Evolving in isolation from the often more aggressive newcomers, the native wildlife often cannot cope with the new pressures. There are many examples of the serious damage that can be caused by introduced species spread by growing global trade.

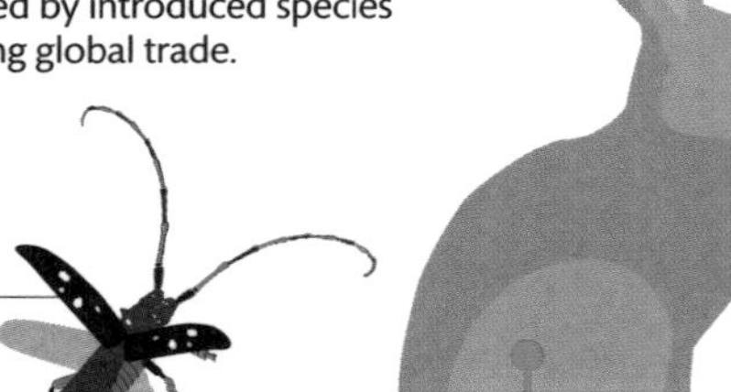
Adult beetles feed on twigs and leaves, larvae burrow deep inside trees
A female rabbit produces 18–30 young each year
Vine can grow up to 26 cm (10 in) per day

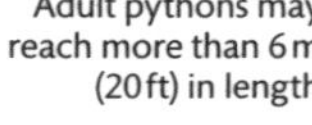
Adult pythons may reach more than 6 m (20 ft) in length

Asian long-horn beetle

Native to China and Korea, these insects have devastated trees in parts of Europe and in the USA – where from 1996–2006 the cost of eradiction attempts was over US$800 million.

European rabbit

Rabbits have changed natural habitats across the world. They breed rapidly – 24 rabbits introduced to Australia in 1894 produced 10 billion by the 1920s. They compete with native species for food.

Burmese python

Originally escaped pets imported from South and Southeast Asia, these huge snakes now threaten rare wildlife across Florida, USA. They predate and outcompete native species.

Kudzu vine

These climbing vines are native to Southeast Asia but are smothering ecosystems from the USA to New Zealand. The vine grows rapidly over plants and trees, creating vast monocultures.

What can we do?

- **Countries must do more to prevent the importation of invasive species.** This can be achieved through effective trade controls, including on certain kinds of garden plants and marine species carried in ships' ballast tanks.

What can I do?

- **Never deliberately release pets or garden plants.** Many of the most damaging alien invaders arrived via this route. Once they are out, it is often impossible to stop them spreading.
- **Take care when disposing of garden waste.**

Every day an estimated **7,000 species are carried around the world** in ship ballast water

Aquatic invasive species

Ocean ships move marine wildlife around the world, both in their ballast tanks that hold seawater, and also attached to the outside of their hulls. Many rich and varied freshwater ecosystems have also been seriously damaged by invasive species. This is one reason why freshwater fish are one of the most threatened animal groups.

Caulpera seaweed

Popular marine aquarium plants, caulpera seaweeds are causing major problems across the Mediterranean, where they smother native seaweed, and invertebrates, causing the decline of many species.

Nile perch

Native to many African rivers, the introduction of these voracious predators into African lakes caused the extinction of several hundred fish species through direct predation and competition for food.

Zebra mussel

These molluscs spread from western Asia during the 1700s and reached the Canadian Great Lakes in the 1980s. They reduce numbers of phytoplankton available to fish larvae and can devastate entire food chains.

Forms dense meadows on seabed, blocking out other marine life

Can grow up to 2 m (6½ ft) in length

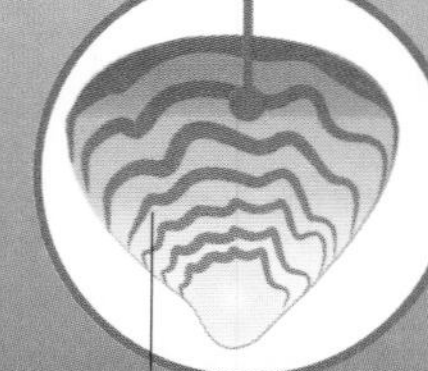

Filters up to 2 litres (4 pints) of water per day

Nature's services

Natural systems and wild species are not only beautiful but provide a wide range of essential and economically valuable benefits. These are sometimes referred to as ecosystem services. They range from flood protection given by forests to the storage of carbon in wetlands, and from the pollination of crops by wild insects to the replenishment of freshwater by wetlands. Often, however, economic growth is achieved at the expense of the health of natural systems. For example, all of our food plants and animals, and many of our medicines, are derived from wild species. By permitting extinctions, we are closing down future opportunities for innovation in food and healthcare. A healthy marine food web depends on plankton – without them, fish stocks would be hugely depleted.

PHOTOSYNTHESIS

Light energy is absorbed by cells in leaf

Oxygen is released as a by-product

Leaf cells absorb carbon dioxide and water

Photosynthesis makes glucose and other foodstuffs for energy and growth

Tourism
Natural habitats, such as beaches, mountains, and forests, are the basis of multi-billion-dollar tourism industries. Access to natural areas improves mental and physical health.

MARINE FOOD CHAIN

Killer whale is top predator

Phytoplankton are at bottom of food chain, harnessing energy from sunlight

Bigger fish prey on smaller species. Small fish feed on plankton

Zooplankton are primary consumers feeding on phytoplankton

Coastal protection
Ecosystems such as mangroves and salt marshes protect coastal areas from inundation by the sea.

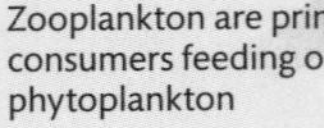

Capture fisheries
"Solar-powered" plankton in the oceans is the basis of a food web that sustains some 90 million tonnes (99 million tons) of fish capture each year. This is the major source of protein for about one billion people.

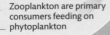

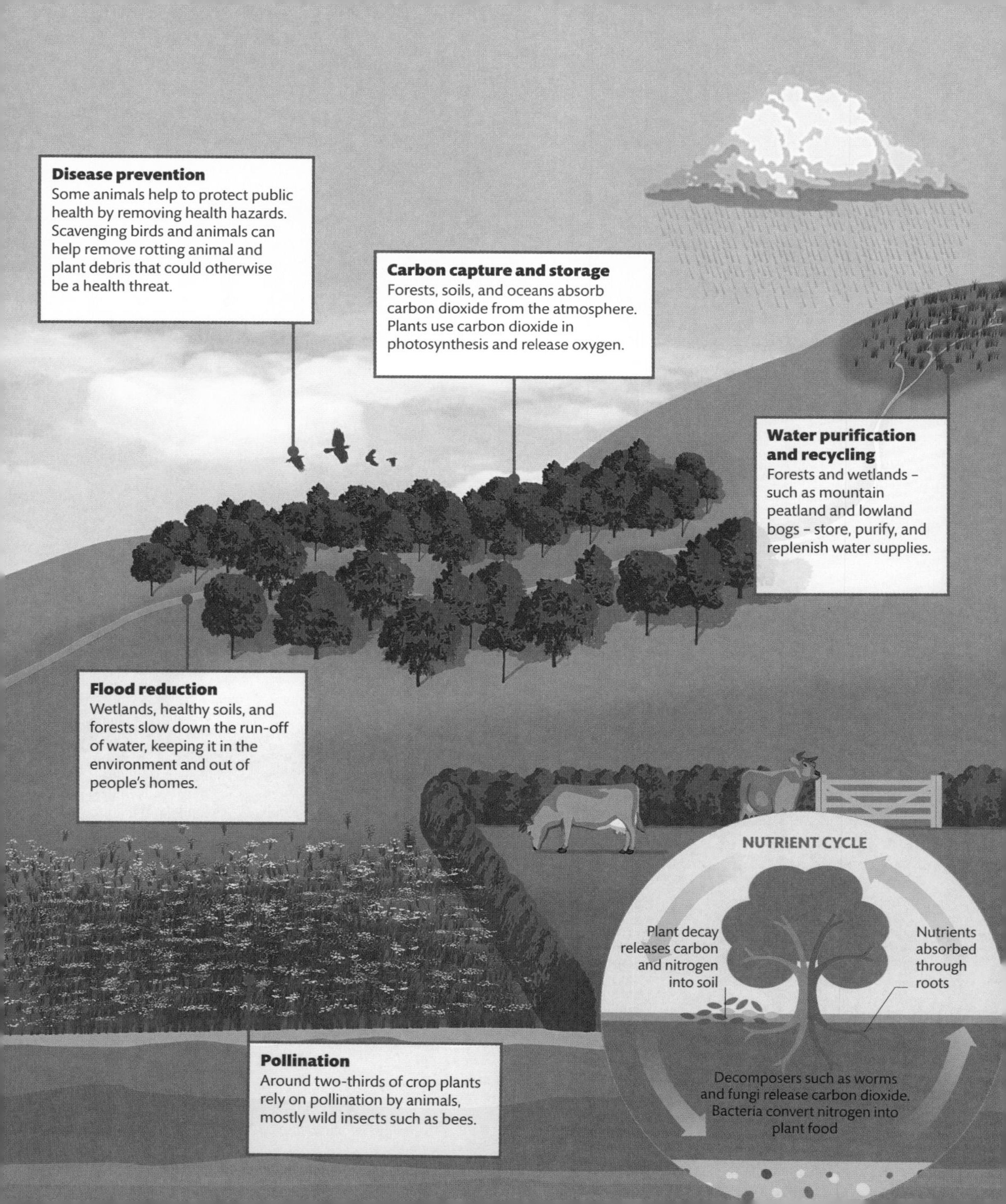

Disease prevention
Some animals help to protect public health by removing health hazards. Scavenging birds and animals can help remove rotting animal and plant debris that could otherwise be a health threat.

Carbon capture and storage
Forests, soils, and oceans absorb carbon dioxide from the atmosphere. Plants use carbon dioxide in photosynthesis and release oxygen.

Water purification and recycling
Forests and wetlands – such as mountain peatland and lowland bogs – store, purify, and replenish water supplies.

Flood reduction
Wetlands, healthy soils, and forests slow down the run-off of water, keeping it in the environment and out of people's homes.

NUTRIENT CYCLE

Plant decay releases carbon and nitrogen into soil

Nutrients absorbed through roots

Decomposers such as worms and fungi release carbon dioxide. Bacteria convert nitrogen into plant food

Pollination
Around two-thirds of crop plants rely on pollination by animals, mostly wild insects such as bees.

Insect pollination

Almost 9 in 10 species of land plants – including most crop varieties – rely on pollination by animals, especially insects, to complete their life cycles. But wild insect populations are declining, posing a risk to food security.

Bees, wasps, hoverflies, butterflies, and beetles are among the insects that pollinate flowers, enabling plants to produce seeds and fruit. Most of the fruits and vegetables we eat rely on insects. In some parts of the world, the loss of wild pollinators has already disrupted food production, forcing farmers to resort to extreme measures, including pollinating plants by hand using paint brushes. Such cases reveal not only the crucial role of pollinators in the food chain but also their huge economic value. Their annual contribution has been estimated to be worth about US$190 billion globally, including US$14.6 billion in the USA and US$600 million in the UK.

SEE ALSO...

› **Nature's services** pp164–65

TYPES OF POLLINATORS

Insect pollination first evolved about 140 million years ago and plays a vital role in ecosystem function. There are several types of pollinators. Some are highly specialized, visiting just one species of plant; others are very general, feeding on a wide range of flowering plants.

Bees

A variety of bees undertake pollination, including bumblebees, solitary bees, mason bees, carpenter bees, and honeybees.

Wasps

Many of the 75,000 wasp species pollinate one particular species of plant. Some live in colonies; others are solitary.

Hoverflies

Adults feed on nectar and pollen while the larvae are aphid predators, making them both pollinators and pest controllers.

Butterflies and moths

These insects use their long proboscis to feed on nectar deep inside flowers, and, in doing so, transfer pollen between blossoms.

Threats to pollinators

In many areas of the world, wild pollinators have undergone drastic decline, mainly as a result of agriculture. Habitat loss due to farming deprives insects of food plants and breeding areas, and many pesticides are toxic to pollinators. In common with other wildlife, many pollinators are also being affected by climate change, as well as other threats, such as housing and infrastructure developments and pollution. Shown here are the principal threats to bees in Europe.

AGRICULTURE
The progressive intensification of farming has led to more and more species disappearing from farmed land. Pesticides have devastated some populations of insect pollinators, while herbicides have killed wild flowers, depriving pollinators of food

LIVESTOCK
More intensive livestock rearing has involved the replacement of traditional hay meadows with silage production. In some countries, such as the UK and Sweden, more than 95% of flower-rich grasslands have been lost, depriving pollinators of vital habitats

POLLUTION

Nitrogen deposition arising primarily from fertilizers causes plant diversity to decline in grasslands, wetlands, and other habitats, depriving pollinators of their food sources

What can we do?

- **Governments could ban the most damaging pesticides**, including the neonicotinoids that are harmful to bumblebees and birds (see p63).
- **Subsidies to farmers** could be paid only on condition that farmers protect or restore pollinator habitats.

What can I do?

- **Grow pollinator-friendly flowering plants** in your garden and leave wilder patches where insects can hibernate and breed.
- **Buy organic fruit and vegetables;** these are produced without any pesticides that can poison pollinators.

The estimated **economic value of bees** and other pollinators **per year** is **US$190 billion**

Urban expansion and infrastructure development reduce wild and semi-wild areas while fragmenting and further isolating those that remain

Heavy rainfalls, droughts, heatwaves, and alterations in the timing of seasons can adversely affect populations of insect pollinators

Sea defences that affect coastal habitats can impact species that are specially adapted to those habitats

CLIMATE CHANGE

FIRE AND FIRE SUPPRESSION

OTHER ECOSYSTEM CHANGES

POLLINATION
Bees and other pollinators transfer pollen from one flower to another, enabling the plants to reproduce

RESIDENTIAL AND COMMERCIAL DEVELOPMENT

RECREATIONAL DISTURBANCES
Tourism in wild or semi-wild areas, such as ski tourism in the Alps, can disturb natural habitats, threatening bees and other pollinators

Fire has its greatest impact on species in drier areas. Land management intended to reduce the fire risk can also reduce plant diversity

MINING AND QUARRYING
Mineral extraction leads to loss of vegetation, but rehabilitated mines and quarries can provide excellent habitats for insects

The importance of bees

Healthy diets include a diverse range of fruit and vegetables. Maintaining a secure supply of such produce into the future will depend on healthy insect populations. Domesticated honeybee hives can play some role, but many crops rely primarily on other species, such as wild bumblebees; in the UK, for example, at least 70 per cent of crop pollination is carried out by wild insects.

Pollinating by hand
In parts of south-western China, the destruction of wild pollinators with pesticides means that fruit growers must pollinate blossoms by hand.

Value of nature

It is often assumed that environmental damage is an inevitable price of progress. However, the loss of free services provided by nature is creating major costs and risks.

Nature provides a wide range of essential services that sustain development. It is possible to estimate the financial value of these, such as the work done by bees in pollinating crops, the importance of coral reefs in protecting coasts from storms, and the role of wetlands and forests in replenishing freshwater. The economic value of natural services is vast and estimated to be worth more than global GDP.

Global GDP

US$66.9 trillion

US$39.7 trillion

Nature's bounty

Work by US environmental economist Robert Costanza and colleagues has revealed the value of nature and how the financial value of ecosystem services changed between 1997 and 2011. A range of valuation methods were used, but this research demonstrates how nature's annual contribution is bigger than world GDP. These findings reveal that the continuing development of human societies depends directly on the health of nature. The more we damage ecosystems, the bigger the costs to human societies in replacing what nature once did for free.

"Nature is our home. Good economics demands we manage it better."

PROFESSOR SIR PATHA DASGUPTA, AUTHOR OF THE UK GOVERNMENT'S 2021 DASGUPTA REVIEW

What can we do?

- **Governments and companies** can gather information on their impact on and their dependence upon natural assets. This information can shape economic decisions to improve, rather than decrease, the health of vital ecosystems.

Value of GDP

While countries seek growth in GDP, the declining health of nature is absent from economic calculations. As ecosystems are destroyed and degraded, the value we get from them is declining.

Global value of nature

Natural systems

All around us ecosystems and wild species help to sustain human welfare. Carbon dioxide is removed from the air by forests, which helps to slow climate change. Wild fish stocks are replenished by solar-powered food webs that begin with plankton, and provide nutrition and jobs. New drugs and crop varieties are being developed with genetic material found in wild species. The contribution of nature is shown in estimates by Costanza and his team.

KEY

FOREST
The economic value of forests is US$16 trillion per year. Forests replenish oxygen, supply water, and are home to most land species.

GRASSLANDS
Different kinds of grasslands are estimated to deliver over US$18 trillion in value, through sustaining most of the world's livestock.

WETLANDS
These help to reduce flood risk, capture carbon, and purify water. Wet ecosystems deliver over US$26 trillion in value.

LAKES AND RIVERS
Our water supplies depend on lakes and rivers being replenished: an annual economic contribution in excess of US$2 trillion per year

CROPLAND
The croplands that grow our food depend on soils supplying nutrients to plants. They provide services worth over US$9 trillion per year.

URBAN
Semi-natural environments in towns and cities provide valuable services. The global value of these per year is over US$2 trillion.

OPEN OCEAN
This global asset provides services worth nearly US$22 trillion per year, including ocean plants that produce much of Earth's oxygen.

COASTAL
Ecosystems lying where sea meets the land provide US$28 trillion-worth of services, such as protection from storms and tourism.

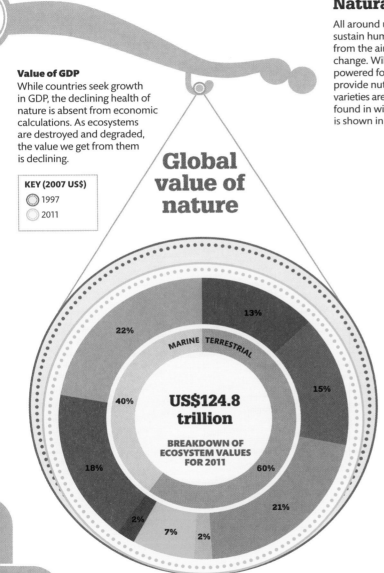

Pandemic risk

Diseases have spread to humans from animals on multiple occasions throughout history, causing major disruptions to societies. Today that risk is increased.

The transmission of disease from animals and to human populations is not new. Diseases with animal origins include, among many others, bubonic plague, rabies, anthrax, Ebola, tuberculosis, yellow fever, measles, and influenza. However, our behaviour in relation to the natural world and wildlife increases the risk of novel diseases jumping from animals to humans. The risk of disease is a key factor underlining the importance for human wellbeing of nurturing a healthy relationship with the natural environment.

What can we do?

- **Governments and companies** can gather information on the impact of their policies on natural assets such as forests and rivers. This informaton can help to shape economic decisions to improve, rather than damage, the health of wildlife and ecosystems.

How environment affects pandemic risk

Our relationships with nature and wildlife can increase the risk of diseases jumping from animals. Outbreaks of the deadly Ebola virus have been associated with deforestation in Africa. The outbreak of severe acute respiratory syndrome (SARS) in 2003–2004 was traced to Guangzhou, China, where live palm civets (catlike tree-dwelling animals) were traded and kept for food. The deadly Nipah virus, which was first found in humans in the late 1990s, originated in bats, then infected pigs, before jumping to people.

Factory farming

Strains of influenza, known as swine flu and bird flu, which can pose serious danger to human life have originated in factory farms. The cramped and stressful conditions enable viruses to mutate and spread, sometimes leading to strains that can infect humans. The cultivation of crops for animal feed used in factory farms contributes to deforestation.

Wildlife trade

Animals taken from the wild and traded for food, skins, medicinal products, or as pets can bring diseases from the wild and into the human world. Despite many laws and policies in place across the world, a massive illegal trade in wildlife continues. In addition to driving many species toward extinction, this trade presents a huge risk to human health.

Deforestation

As forests give way to farms and mines, animal populations are concentrated into remaining patches of forests. When people move into previously undeveloped areas, so contact between humans, wildlife, and livestock is increased. The risk rises as animals venture out of the remaining forest to find food among crops and as people inhabit the forest edge.

Hunting

Hunting wild animals for food can lead to the transfer of diseases to people. HIV AIDS originated from the bushmeat trade, in which apes and monkeys are killed to supply demand for their meat. Blood contact between the hunters and the hunted caused this deadly virus to infect humans. This disease spread worldwide, causing millions to die.

Bat
The genetic sequence of SARS-CoV-2 (Covid-19) suggests it originated in bats, which may have been caught and brought to markets.

Pangolin
Pangolins, traded for food and medicine, come into contact with other animals, including live bats, in markets, where they may have contracted the virus.

Human
The virus is believed to have multiplied in pangolins (or another species), and then infected humans, following contact in markets.

Spread of animal viruses to humans

In late 2019, a new virus outbreak was identified in Wuhan, China. Named SARS-CoV-2, it was found to be the cause of a severe respiratory disease in humans, called COVID-19. The origin of the infection was traced to a so-called wet market, where different species of live animals, including bats and pangolins, were sold for food. People in close contact with pangolins are likely to have been the first to contract the disease, which then spread more widely.

Around **60 per cent** of existing and emerging infectious diseases **originate in animals,** with most of those coming from **wild animals**

Global transmission
In our interconnected world, infectious diseases can spread between countries very quickly. Less than four months after COVID-19 was first identified in China as a new virus, the World Health Organization declared a global pandemic.

Consequences of the pandemic

The first cases of COVID-19 were reported in Wuhan in China in late 2019. Tens of millions of people were infected and millions died. On top of the direct health effects, were a range of other consequences.

Serious health, societal, and economic consequences followed in the wake of the pandemic. Businesses closed, global trade was impacted, governments enacted restrictive new laws, travel patterns were drastically altered, and public health systems in some countries were brought to breaking point. Costly programmes to identify treatments and vaccines were rapidly instigated. Many families were affected by bereavement and stress and mental health challenges. Many countries were unprepared for the impact of the virus and even when infections became widespread, some governments failed to take the necessary actions.

The costs of the pandemic

The financial amd political costs of dealing with the factors that increased the risk of pandemics, such as halting deforestation or controlling the wildlife trade, had previously been seen as too great to conside. However, the vast costs incurred by governments of dealing with the consequences of the pandemic proved far larger than the sums that might have been spent on measures to prevent pandemics of this kind.

Economic growth

The global economic impact of COVID-19 will be huge. One estimate placed the total cost over five years at nearly US$27 trillion. Business restrictions led to a slow-down in economic activity – and for some countries the contraction was the biggest recorded for centuries. At the same time unemployment caused tax revenues to governments to fall and health and welfare costs vastly increased.

Wildlife hunting

In some African countries wildlife tourism guides lost their jobs because people could not travel. With less money to buy food, they were forced to hunt instead, including the animals that they were previously hired to protect.

Social disruption
As COVID-19 spread, countries across the world were forced to introduce often unpopular restrictions on personal freedoms, such as the compulsory wearing of face masks.

Carbon dioxide emissions

During April 2020 the pandemic caused an estimated 17 per cent cut in CO_2 emissions across the world because of reduced travel and the economic slow-down. In 2020, the reduction was estimated at about 7 per cent. The International Energy Agency predicted that in 2021 global energy demand would rise by 4.1 per cent following a post-pandemic economic rebound.

Hunger

Hunger increased in countries where loss of income meant people could no longer afford enough food. Food prices in some countries also rose. The UN projected an 82 per cent increase in the number facing crisis-level hunger in 2020 over 2019.

Travel disruption

Many countries saw a severe and dramatic reduction in plane flights, with business and holiday traffic decimated in a matter of weeks.

Death toll

By the start of June 2021, there were an estimated 3.3 million dead from COVID-19. The number is likely to have been higher, however, due to the under-recording of deaths in remote rural areas in developing countries. Some countries which initially managed to keep infections low, later struggled to contain the pandemic.

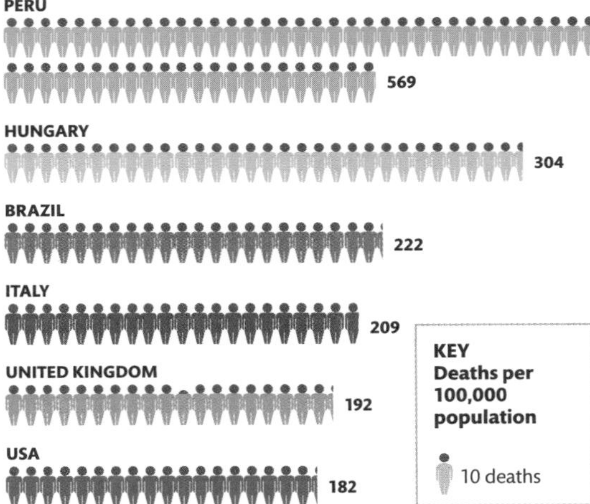

The majority of **COVID-19 deaths** were among people over the **age of 75 years**

GREEN RECOVERY

Many leaders in business, politics, and social movements across the world have called for a green recovery from the pandemic. This would include investing in clean technologies and the restoration of the natural world as a means of creating jobs and stimulating the economy as the world pandemic subsides. While some have suggested that the pandemic might have helped the environment by cutting emissions and reducing economic demand, it is clear that this is neither a sufficient, long-lasting, or widely supported route to dealing with environmental challenges.

"**The core values** that underpin sustainable development – **interdependence, empathy, equity, personal responsibility and intergenerational justice** – are the only foundation upon which any **viable vision of a better world** can possibly be constructed."

SIR JONATHON PORRITT, BRITISH ENVIRONMENTALIST AND WRITER

The Great Acceleration

What's the global plan?

Shaping the future

3 BENDING THE CURVES

A wide range of initiatives is in place to address interrelated global challenges, but if we are to achieve a secure and sustainable future, far more will be needed.

The Great Acceleration

The pressures exerted by humankind on planet Earth have led to fundamental changes to the atmosphere, ecosystems, and biodiversity while depleting many resources. Further population and economic growth are driving the demand that is behind continuing changes, many of which are interconnected. The scale of human activity is so big as to become the most influential factor shaping life on Earth. Scientists believe we have entered a new geological era – the Anthropocene – a period in which people have become a defining global force.

A new era: the Anthropocene

The point at which the Anthropocene began is subject to debate. Some suggest it began during the Pleistocene, up to 50,000 years ago, when humans caused the extinction of many large mammals. Others suggest it coincides with the rise of agriculture. There is a strong argument for the industrial revolution as the point from which to start the new epoch, since it ushered in an unprecedented global impact on the planet. Equally, some argue it began when the first atomic bomb was detonated, leaving a global radioactive human fingerprint. There is increasing agreement, however, that the 1950s is the best place to mark the start of the Anthropocene. This was the start of a unique period, called the Great Acceleration, when many human activities reached take-off points and sharply accelerated towards the end of the century.

50,000 YEARS AGO
Groups of hunter-gatherers target large mammals for food and other resources including skins and bones.
Although climate changes that accompanied the end of the last Ice Age played a part, it has been estimated that about two-thirds of the many large mammal extinctions that took place in this period were caused by humans.

8,000 YEARS AGO
The near simultaneous rise of agriculture and cities marked a sudden change in human impacts.
Hunter-gatherer societies lived close to nature in the ecosystems they depended upon. Farmers feeding urban populations made fundamental changes to their environment, including forest clearance, which caused carbon dioxide (CO_2) levels to rise, while the builders of cities relied on systematic large-scale resource extraction.

5,000 – 500 YEARS AGO
Soil changes created by human activity spread widely across the world with the rise of agriculture.
Some changes were deliberate and aimed at improving soil quality. Other impacts were inadvertent and led to soils being damaged to the point where they stopped producing crops.

1610
A drop in atmospheric CO_2 concentration coincides with forest regrowth.
The mass mortality of indigenous peoples in tropical rainforest regions, caused by diseases and slavery brought by newly arrived Europeans, meant fields reverted to forests, which removed CO_2 from the air.

Rising trends

When researchers plotted various trends reflecting rising human demands and impacts they expected the curves to begin going up sharply from the start of the industrial age, during the 1700s or 1800s. They found, however, that all these and many other trends really took off during the middle of the 20th century. The Great Acceleration that began in the 1950s and continues today is perhaps the correct point from which to mark the start of the Anthropocene.

"It is difficult to overestimate the **scale and speed of change.** In a single lifetime, **humanity** has become a **planetary-scale geological force.**"

WILL STEFFEN, EXECUTIVE DIRECTOR OF THE INTERNATIONAL GEOSPHERE-BIOSPHERE PROGRAMME

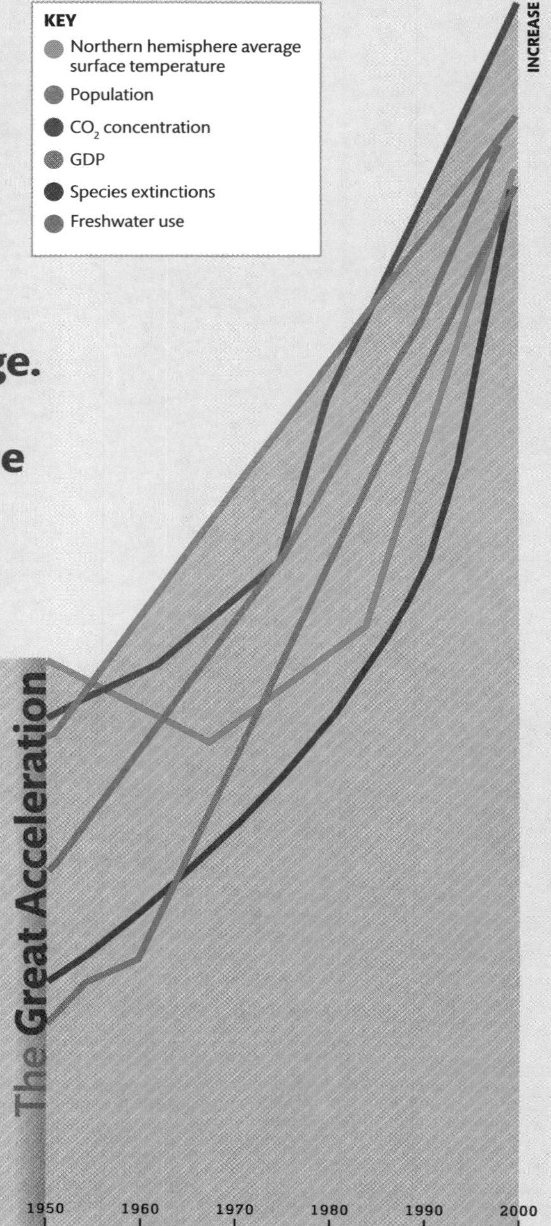

LATE 1700s
The industrial revolution begins in England but soon spreads across Europe and to North America.
The large-scale combustion of fossil fuels begins and there is a sharp increase in demand for other natural resources. Industrialized farming follows in its wake. It took more than 200 years for industrialized development to spread across the globe.

1950
The Great Acceleration: the beginning of rapid growth in many areas.
Following the first nuclear bomb detonation, the Great Acceleration marks the rise of truly global impacts caused by people on planet Earth. As well as leaving a radioactive marker in sediments across the world, climate change, ocean acidification, widespread soil damage, and a mass extinction of species accompany the sharp increase in human influence.

Planetary boundaries

The degradation of the Earth's systems poses increasing risk to human societies. Scientists have identified a number of planetary "boundaries" that if breached could lead to potentially disastrous consequences.

Crossing boundaries

An international team led by scientists at the Stockholm Resilience Centre has set out nine "planetary boundaries" that are believed to be key to the health of our planet. The boundaries relate to global trends, including climate change, ozone depletion, ocean acidification, freshwater use, and biological diversity. The colours depicted here represent the level of risk for each area. Green indicates that to date the risk falls below the boundary – in other words, not presently a globally systemic threat. Yellow is the zone of uncertainty where risk is increasing. Red has gone beyond uncertainty and shows a high risk. Grey is an aspect that has not yet been quantified.

EARTH BUDGET

Human demand is now far larger than what the Earth can indefinitely sustain. Many large economies on use more resources than can be provided within their own borders. For example, Japan needs five times its own area to sustain current consumption. China and the UK are also among countries demanding more than can be provided from their own territory.

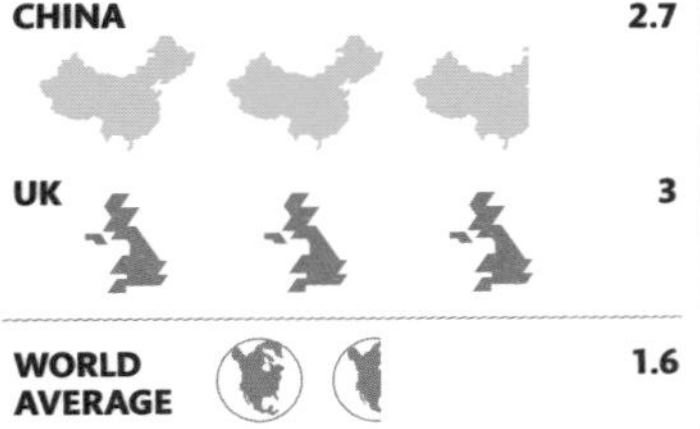

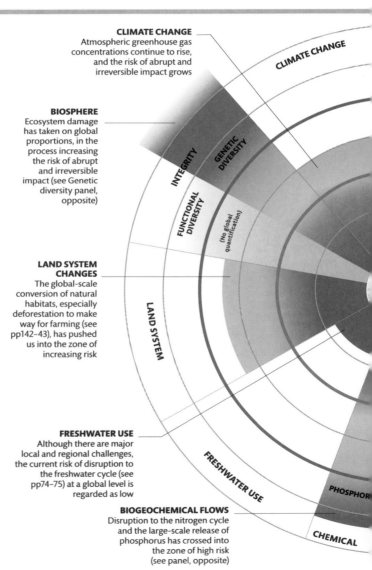

CLIMATE CHANGE
Atmospheric greenhouse gas concentrations continue to rise, and the risk of abrupt and irreversible impact grows

BIOSPHERE
Ecosystem damage has taken on global proportions, in the process increasing the risk of abrupt and irreversible impact (see Genetic diversity panel, opposite)

LAND SYSTEM CHANGES
The global-scale conversion of natural habitats, especially deforestation to make way for farming (see pp142–43), has pushed us into the zone of increasing risk

FRESHWATER USE
Although there are major local and regional challenges, the current risk of disruption to the freshwater cycle (see pp74–75) at a global level is regarded as low

BIOGEOCHEMICAL FLOWS
Disruption to the nitrogen cycle and the large-scale release of phosphorus has crossed into the zone of high risk (see panel, opposite)

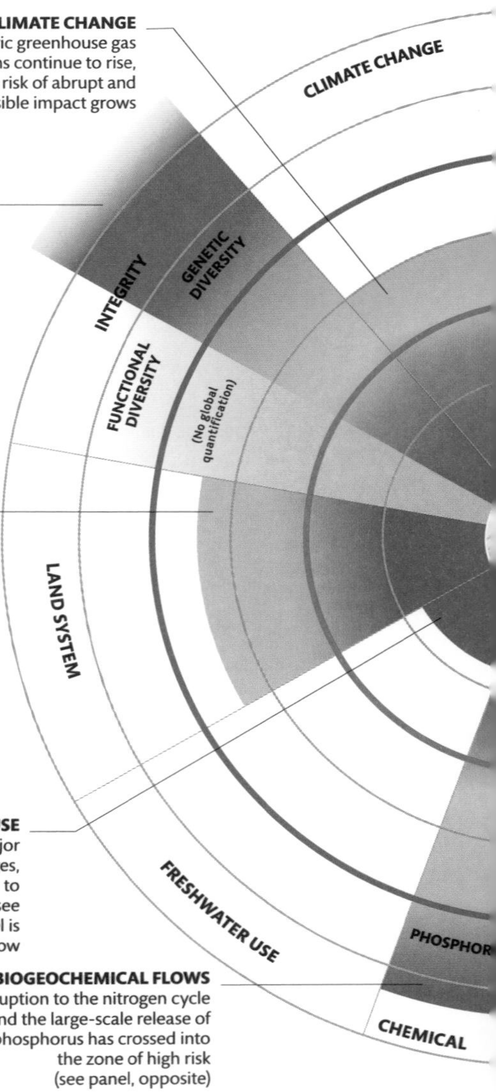

CLIMATE CHANGE
GENETIC DIVERSITY
INTEGRITY
FUNCTIONAL DIVERSITY
(No global quantification)
LAND SYSTEM
FRESHWATER USE
PHOSPHOR
CHEMICAL

It is important to identify those planetary pressures that have become most acute and pose potentially catastrophic risks to humankind. This can help us to prepare for significant change and prioritize resources towards meeting the most pressing challenges. The nine key areas shown here relate to global changes. In many places, local changes are already into the zone of high risk.

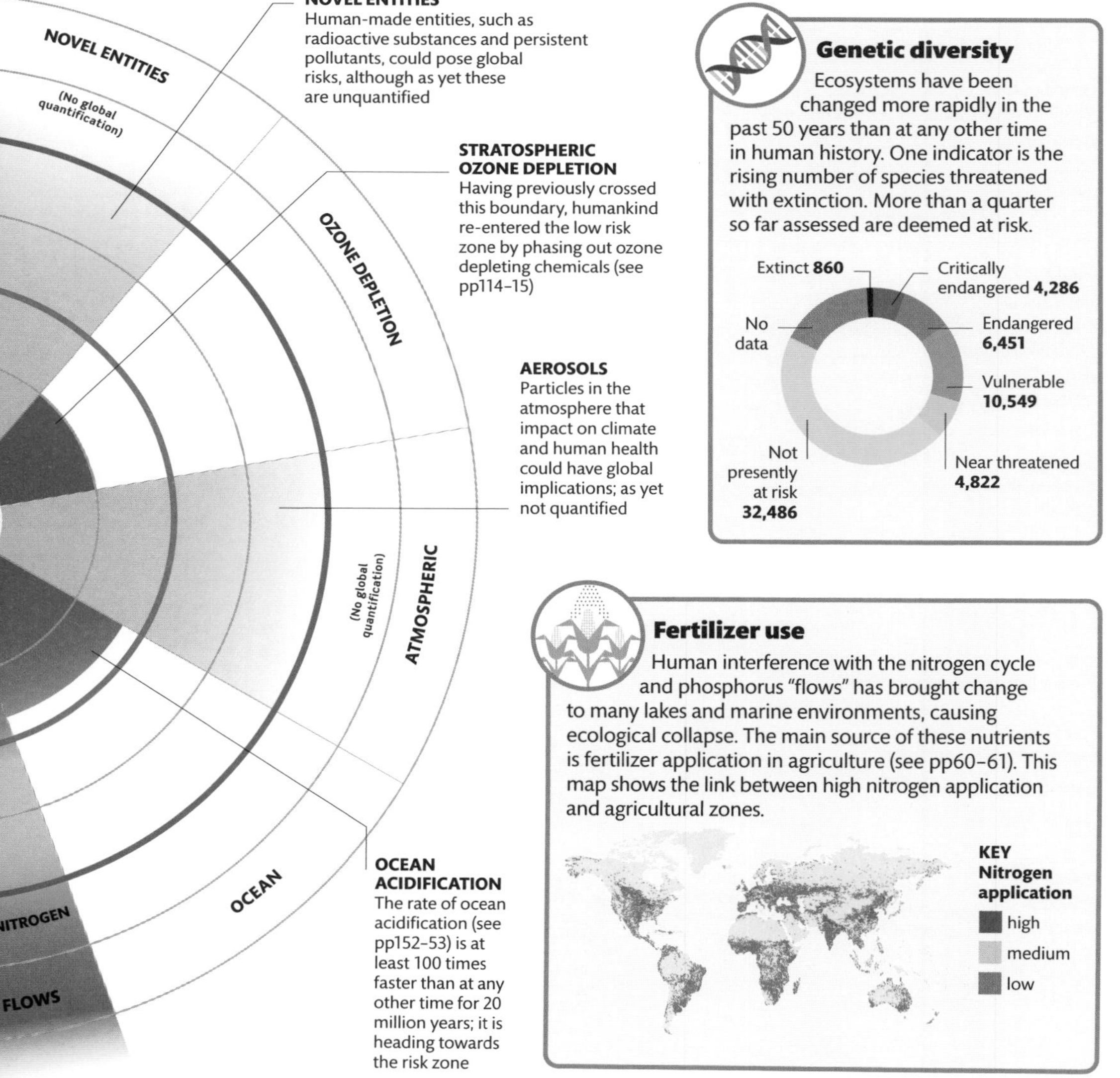

NOVEL ENTITIES
(No global quantification)
OZONE DEPLETION
(No global quantification)
ATMOSPHERIC
OCEAN
NITROGEN
FLOWS
NOVEL ENTITIES
Human-made entities, such as radioactive substances and persistent pollutants, could pose global risks, although as yet these are unquantified
STRATOSPHERIC OZONE DEPLETION
Having previously crossed this boundary, humankind re-entered the low risk zone by phasing out ozone depleting chemicals (see pp114–15)
AEROSOLS
Particles in the atmosphere that impact on climate and human health could have global implications; as yet not quantified
OCEAN ACIDIFICATION
The rate of ocean acidification (see pp152–53) is at least 100 times faster than at any other time for 20 million years; it is heading towards the risk zone
Genetic diversity
Ecosystems have been changed more rapidly in the past 50 years than at any other time in human history. One indicator is the rising number of species threatened with extinction. More than a quarter so far assessed are deemed at risk.
Extinct 860
Critically endangered 4,286
No data
Endangered 6,451
Vulnerable 10,549
Not presently at risk 32,486
Near threatened 4,822
Fertilizer use
Human interference with the nitrogen cycle and phosphorus "flows" has brought change to many lakes and marine environments, causing ecological collapse. The main source of these nutrients is fertilizer application in agriculture (see pp60–61). This map shows the link between high nitrogen application and agricultural zones.
KEY
Nitrogen application
high
medium
low

Interconnected pressures

Rising demand for food, energy, and water presents great challenges, but less obvious are the connections between them. Energy and water produce food, water produces energy, and energy cleans and supplies water.

During 2008, food prices rose significantly, increasing the number of hungry people in the world by an estimated 100 million. This sparked social unrest and led many countries to restrict exports of staple foods. Two of the main reasons for this situation were the unprecedented high price of oil and gas and the droughts affecting major food-producing areas. The future security of human societies depends on finding solutions that recognize the clear links between food, water, and energy. The avoidance of waste and the efficient use of energy, food, and water are essential.

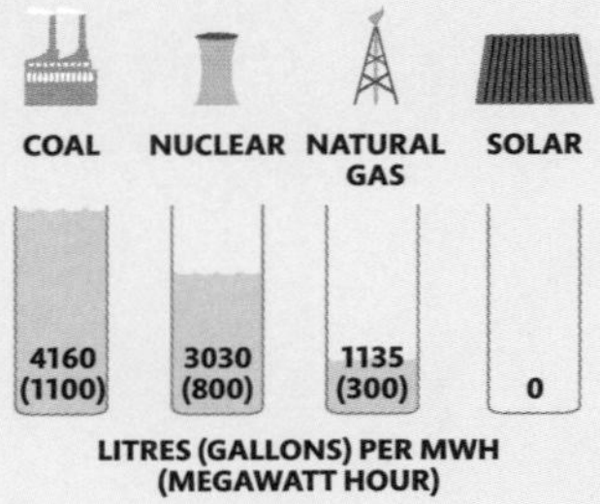

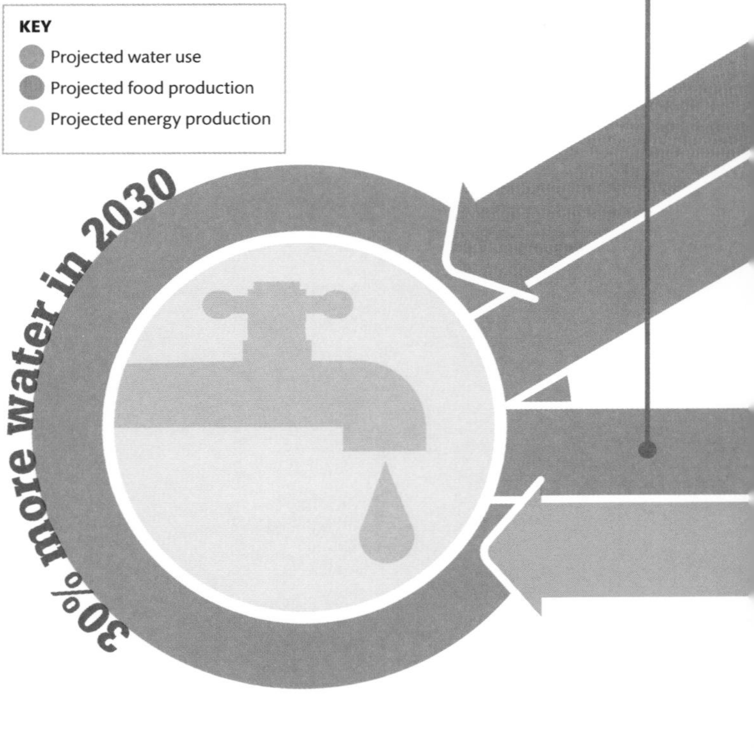

Linked demands

It was estimated in 2014 that by 2030 the world would need 30 per cent more water, 40 per cent more energy, and 50 per cent more food. Meeting these rising demands will be individually challenging, but the pressures emerging between them have been described as potentially creating a "perfect storm". This graphic shows some of the implications of the growing demand for food, energy, and water, and how the increased consumption of one has implications for the others.

Water for power

Water is vital for many different sources of power production, especially coal and nuclear, where it is used for cooling. Renewable technologies, such as solar photovoltaic, do not need any water.

COAL	NUCLEAR	NATURAL GAS	SOLAR
4160 (1100)	3030 (800)	1135 (300)	0

LITRES (GALLONS) PER MWH (MEGAWATT HOUR)

KEY

- Projected water use
- Projected food production
- Projected energy production

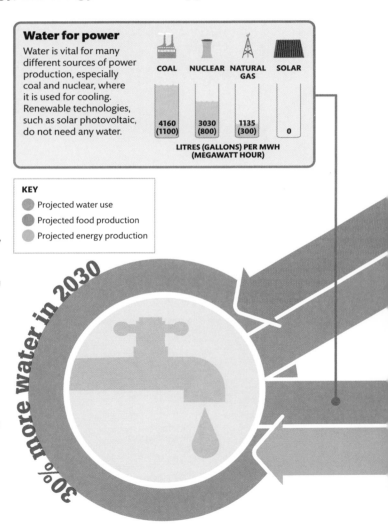

30% more water in 2030

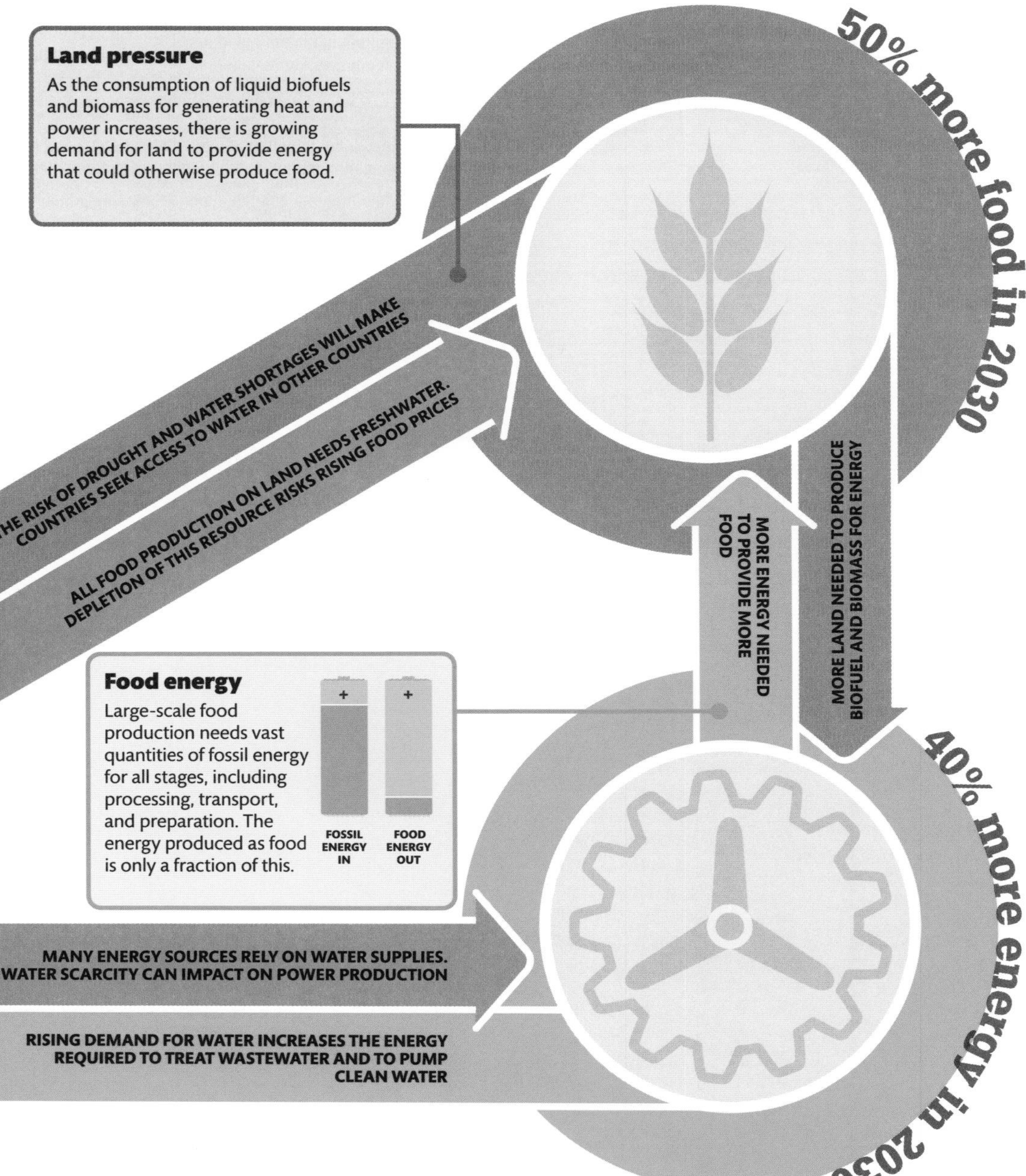
Land pressure
As the consumption of liquid biofuels and biomass for generating heat and power increases, there is growing demand for land to provide energy that could otherwise produce food.
50% more food in 2030
THE RISK OF DROUGHT AND WATER SHORTAGES WILL MAKE COUNTRIES SEEK ACCESS TO WATER IN OTHER COUNTRIES
ALL FOOD PRODUCTION ON LAND NEEDS FRESHWATER. DEPLETION OF THIS RESOURCE RISKS RISING FOOD PRICES
MORE ENERGY NEEDED TO PROVIDE MORE FOOD
MORE LAND NEEDED TO PRODUCE BIOFUEL AND BIOMASS FOR ENERGY
Food energy
Large-scale food production needs vast quantities of fossil energy for all stages, including processing, transport, and preparation. The energy produced as food is only a fraction of this.
FOSSIL ENERGY IN
FOOD ENERGY OUT
40% more energy in 2030
MANY ENERGY SOURCES RELY ON WATER SUPPLIES. WATER SCARCITY CAN IMPACT ON POWER PRODUCTION
RISING DEMAND FOR WATER INCREASES THE ENERGY REQUIRED TO TREAT WASTEWATER AND TO PUMP CLEAN WATER

What's the global plan?

Recognizing the limited ability of individual countries to solve many environmental problems, intensive efforts have been devoted towards the negotiation and implementation of various multilateral environmental agreements (MEAs). These are formal legal accords between countries to manage collective challenges that no one country can meet on its own. Countries signing multilateral agreements undertake to implement commonly agreed rules and meet targets linked with different environmental challenges.

Rise in MEAs

During the last century, the number of international environmental treaties, protocols, and other agreements has grown, especially during the 1970s, 80s, and 90s. While some agreements have been highly successful in galvanizing coordinated responses, many have struggled to meet their aims. Some have attracted support steadily over time, but others have achieved very rapid sign up from countries. For example, when countries realized the serious risks and threats posed by the loss of the Earth's natural diversity, support for the Convention on Biological Diversity quickly grew.

KEY
World Heritage Convention
Adopted at UNESCO's General Conference in 1972 to stem threats to natural and cultural heritage sites.
CITES
Agreement adopted in 1973 and entered into force in 1975. Aims to protect species that are traded.
Vienna/Montreal
Entered into force in 1988 to protect the Earth's ozone layer.
Basel
Adopted in 1989 and entered into force in 1992 to control the international shipment of hazardous wastes and their disposal.
UNFCCC
United Nations Framework Convention on Climate Change (UNFCCC) and Kyoto Protocol. Convention agreed in 1992 and Protocol in 1997. Paris Agreement 2015.
CBD
United Nations Convention on Biological Diversity (CBD). Agreed at the 1992 Rio de Janeiro Earth Summit. The US refused to sign.
1988
World reacts with unprecedented speed to scale up action to save the ozone layer with the Vienna/ Montreal agreements.
1972
1975
1980
1985
YEAR

MULTILATERAL ENVIRONMENTAL AGREEMENTS

During the last century, hundreds of new international environmental agreements have been reached. Most are technical amendments to existing plans, others are major new treaties. Over time, as more and more MEAs have been adopted, the rate of new ones being agreed has gone down. It is not for want of new agreements that the world struggles to make progress, but more the effective implementation of what's already there.

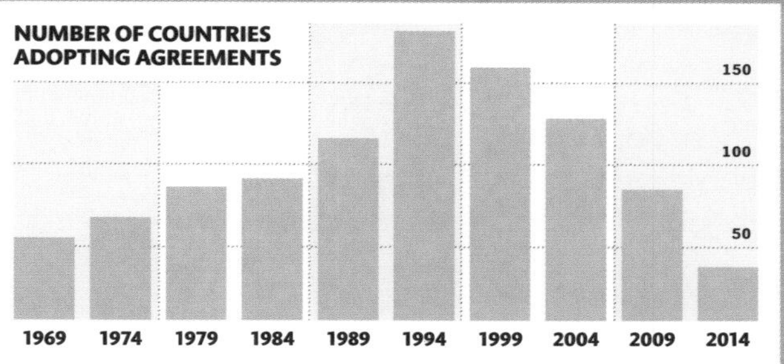

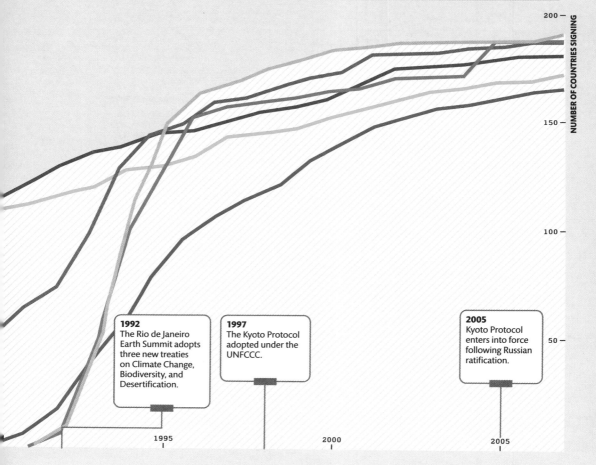

200
150
100
50
NUMBER OF COUNTRIES SIGNING

1992
The Rio de Janeiro Earth Summit adopts three new treaties on Climate Change, Biodiversity, and Desertification.

1997
The Kyoto Protocol adopted under the UNFCCC.

2005
Kyoto Protocol enters into force following Russian ratification.

1995
2000
2005

Sustainable development goals

In 2015, the Millennium Development Goals set in 2000 expired. New action was needed to set out a framework for meeting environment and development challenges until 2030 and to lay the foundations for a more secure future.

The world first committed to the goal of sustainable development at the Earth Summit in Rio in 1992, but societies everywhere struggled to embed its central idea of meeting the needs of the present without compromising the needs of future generations. Instead, economic growth and progress towards social goals were promoted at the expense of environmental assets and climatic stability. In 2012, governments – supported by campaign groups and major international companies – agreed to establish a new set of goals. In 2015, following a period of intense negotiations, the United Nations General Assembly adopted the Sustainable Development Goals (SDGs), to be achieved by 2030. However, by the end of 2019, progress had been patchy toward meeting these goals in relation to many of the key indicators set. The world is doing better on social and economic goals than on environmental ones.

GOOD HEALTH
There has been strong progress with improved access to healthcare in many parts of the world. But sub-Saharan Africa lags behind.

NO POVERTY
Poverty reduction strategies have been relatively successful in South and East Asia, but in Africa and Latin America major challenges remain.

QUALITY EDUCATION
Outside developed countries only sub-Saharan Africa has seen decisive progress, albeit starting from a lower level than other regions.

RESPONSIBLE CONSUMPTION
Attempts to delink material consumption from economic growth have stalled, except in already-rich countries.

CLEAN WATER
Progress in central and southern Asia and developed countries is set against limited progress across Africa and Latin America.

INDUSTRY AND INNOVATION
East and Southeast Asia have shown strong progress, but in Latin America there has been a deterioration since 2015.

ZERO HUNGER
The world has struggled to make progress toward improved nutrition and food security with many millions of people still hungry.

CITIES AND COMMUNITIES
With the aim of reducing the numbers living in city slums, developed countries continued to improve. North Africa and western Asia performed worst.

KEY
Progress toward the SDGs (late 2019)
- Substantial progress
- Limited progress
- Fair progress but acceleration needed
- Deterioration

ACCESS TO ENERGY
Much progress has been made in the provision of electricity, with the exception of North Africa and western Asia. But most increase has been achieved using fossil fuels.

LIFE BELOW WATER
Rapid and substantial progress has been achieved towards the establishment of protected areas of coast and ocean. Concern remains about the actual protection provided.

PEACE AND JUSTICE
Positive progress at the global level contrasts with the deterioration since 2015 seen in North Africa and western Asia.

GROWTH
Strong economic growth has continued since 2015 in many regions, but in some the situation has worsened, particularly in North Africa and western Asia.

GENDER EQUALITY
Positive progress has been made in Latin America and developed countries towards more women in politics. Little change in North Africa and South and central Asia.

PARTNERSHIPS
Sustainable development requires financial input, but at the global level there has been no real improvement in the funding available.

REDUCED INEQUALITY
With the exception of East and Southeast Asia, there has been no overall progress on this measure.

CLIMATE CHANGE
Global efforts to reduce emissions quickly in line with the Paris Agreement have thus far largely failed, although some countries are making progress.

LIFE ON LAND
The trends are either negative or at best without positive progress. The mass extinction of animal and plant species continues largely unabated.

What are the goals?

The 17 SDGs focus on discrete challenges but are all linked. They are concerned with human well-being, envisaging a world free of poverty and hunger, where everyone has access to education, healthcare, and social protection, and access to affordable and sustainable energy. They also address human rights and human dignity. The goals are designed to build a more just, equitable, tolerant, and socially inclusive world. Above all, they are concerned with sustainability and the building of a world in which every country enjoys inclusive and sustainable economic growth, and decent work for all, while also protecting the environment and conserving biodiversity.

What can we do?

- **Urge governments** across the world to adopt ambitious plans for full implementation of the new Sustainable Development Goals.

What can I do?

- **When buying products and services** from international companies, choose those whose policies support the achievement of the goals.

Climate challenges

In 1992 at the Rio de Janeiro Earth Summit, countries across the world adopted The United Nations Framework Convention on Climate Change. Its aim was to avoid dangerous changes to the Earth's climate system.

Since 1992 targets have been set to limit global warming, notably the 2015 Paris Agreement, which says overall increase in global temperature by 2100 needs to be limited to below 2°C (3.6°F) compared with the temperature before 1850, and if possible not to go above 1.5°C (2.7°F). The world is not on track to reach this goal. In 2019, the world recorded the highest ever emissions level, with 59 billion tonnes (65 billion tons) of carbon dioxide equivalent released into the atmosphere.

Failing policies

Overall, current policies place the world on a path that might lead to up to nearly 4°C (7.2°F) of global average temperature increase, which would risk catastrophic consequences. Even this level of increase might prove to be an optimistic assessment, however, considering that most countries have a history of failing to meet environmental goals.

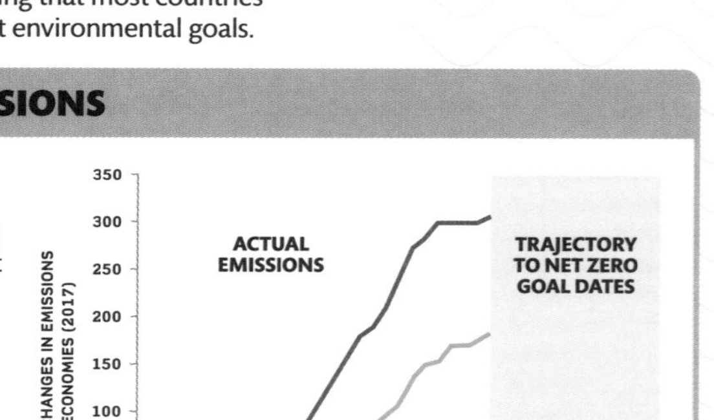

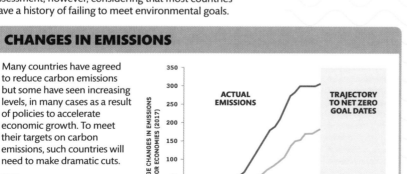

CHANGES IN EMISSIONS

Many countries have agreed to reduce carbon emissions but some have seen increasing levels, in many cases as a result of policies to accelerate economic growth. To meet their targets on carbon emissions, such countries will need to make dramatic cuts.

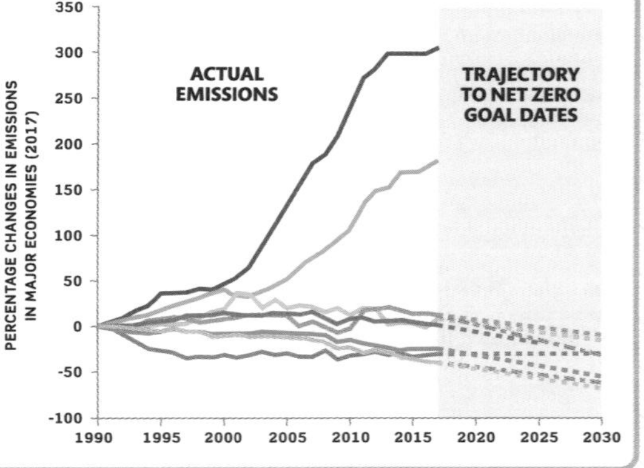

KEY

- Australia
- China*
- EU
- India*
- Japan
- Russia
- UK
- USA

*China and India do not have percentage reduction goals

Countries need to increase their climate targets by **80 per cent more** than they pledged under the **Paris Agreement**

Biodiversity targets

Following warnings about the rapid disappearance of species and ecosystems, in 1992 the Convention on Biological Diversity was first agreed at the United Nations Rio de Janeiro Earth Summit.

This treaty sets out to conserve biological diversity, achieve its sustainable use, and ensure the benefits arising from nature are fairly shared. In 2010, a series of 20 targets were agreed that it was hoped would, by 2020, have put the world on a path to saving the intricate web of life on the planet upon which human societies and all of nature depends. Following the adoption of these targets, however, governments around the world did not prioritize their implementation and in 2020 a review of progress found that at the global level that not one of the targets had been reached.

Portfolio of key actions

The loss of nature is not the result of any single factor, but a combination of pressures that require a deeper level of action to address than has hitherto been delivered by countries across the world. Whereas it was once assumed that setting aside natural areas in reserves and parks would be sufficient to stem the decline of nature, we know from research that, while vital, these measures on their own are not sufficient to halt or redress the decline of biodiversity. In particular, new economic policies that fully reflect and embrace the value of nature to sustainable development and human health are also needed.

SEE ALSO...

- **Forest clearance** pp142–43
- **Biodiversity hotspots** pp160–61
- **The value of nature** pp168–69
- **Restoring the future** pp204–05

The price of progress

The destruction of nature has often been seen as "the price of progress". Research confirms, however, that the opposite is the case and that healthy natural systems are essential for economic development.

Reducing other drivers

Housing and infrastructure impact on biodiversity and must be planned in ways that enhance natural areas

Reduced consumption

Reducing demand for natural resources will be vital in reducing the pressures upon natural systems

Climate change action

Reducing greenhouse gas emissions to slow climate change is essential for protecting nature and conserving biodiversity

Conservation/ Restoration

Alongside conserving wild areas that remain, it is vital to restore much of what's been lost

Sustainable production

Farming, forestry, and fishing all impact on nature, but can be managed in ways that sustain biodiversity

AICHI TARGETS MISSED

The adoption of the 20 so-called Aichi Targets at a meeting of the Convention on Biological Diversity in Japan in 2010 was a major landmark. The targets included action to reduce the rate of biodiversity loss, prevent extinctions, reduce pollution, restore ecosystems, and to secure more financial support for conservation. The subsequent failure to meet any of these targets globally marks a major setback and also provides a wake-up call to the global community.

Back from the brink
When pressures on wildlife are alleviated, such as by reductions in pollution or control of hunting, then species can make comebacks, as the wolf has done across parts of Europe.

Transition pathways

The world has reached the beginning of a mass extinction of species and the loss and degradation of ecosystems on a global scale. However, it is still not too late to avoid the catastrophic loss of species and habitats. During the coming decades, if countries prioritize the necessary actions, including through the restoration of severely damaged ecosystems, transitions in key areas could reduce pressures and lead to the recovery of nature.

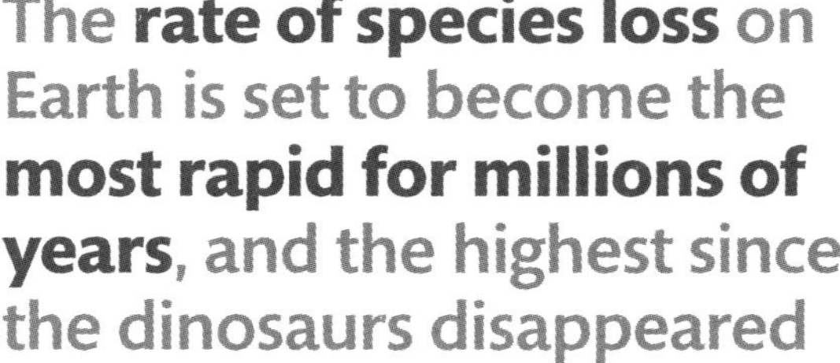

The **rate of species loss** on Earth is set to become the **most rapid for millions of years**, and the highest since the dinosaurs disappeared

Land and forests Conserving intact ecosystems, restoring damaged ones, reversing degradation, and employing landscape-level planning to avoid further damage. Well-conserved habitats retain biodiversity and provide benefit to people.

Freshwater Guaranteeing water flows required by nature and people, improving water quality, protecting critical habitats, controlling invasive species, and safeguarding connectivity to allow the recovery of freshwater systems.

Fisheries and oceans Protecting and restoring marine and coastal ecosystems, rebuilding fisheries, and managing aquaculture and other uses of the oceans to ensure sustainability, and to enhance food security and livelihoods.

Sustainable agriculture Promoting innovative farming systems that enhance productivity while minimizing negative impacts on biodiversity. Such approaches recognize the role of biodiversity in sustaining human food security.

Food systems Promoting sustainable and healthy diets with diverse, mostly plant-based foods, and reduced consumption of meat and fish. Dramatic cuts in food waste reduce demand for land, water, and agricultural chemicals.

Cities and infrastructure Installing "green infrastructure", making space for nature within urban areas to improve the health and quality of life for citizens and to reduce the environmental footprint of cities and built infrastructure.

Climate action Harnessing nature-based solutions to capture carbon from the atmosphere while enabling human communities to adapt to climate change impacts such as floods, droughts, and sea level rise.

One health Conserving and restoring rural and urban environments to promote healthy ecosystems and healthy people. Address the common drivers of biodiversity loss, disease risk (including pandemics), and ill-health.

Nature's spaces

The last 50 years has seen a huge increase in the number of national parks, nature reserves, and other protected areas. While this is a positive trend, there are still many challenges to overcome.

Investment in large, high-quality, and connected areas of natural habitat on land and in coastal and marine areas is vital to minimize the extinction of wild species. In 2010, world governments pledged to increase protected areas as part of the Aichi Biodiversity Targets. However, this will not be enough on its own. Other steps – for example, sustainable farming, enforcement of anti-poaching laws, pollution prevention, and effective action on climate change – are critical for nurturing nature's spaces. Protected areas must also be managed effectively. One recent survey found that only 24 per cent were under "sound management". Experts also conclude that the current protection is insufficient to safeguard the full range of species and ecosystems. Little of the open ocean is protected, and habitats including tropical coral reefs, sea-grass beds, and peat lands need particular attention.

Growth of protected areas

Since 1962, the number of protected sites has increased more than 20-fold globally. By the end of 2020, governments had committed to protecting about 10 per cent of the global ocean and 17 per cent of land and inland waters. Growth in protected areas has been fastest in the marine environment, with almost ten times more ocean protected in 2020 than was the case in 2000.

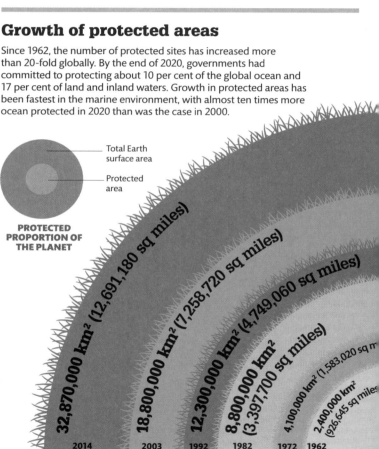

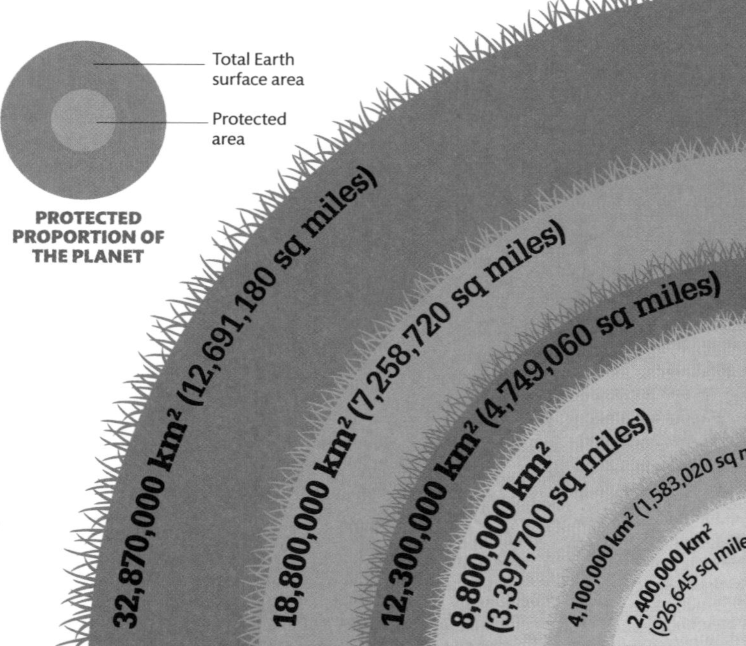

SEE ALSO...

› **Nature's services** pp164–65

› **The value of nature** pp168–69

Protection timeline
The legal protection of land for conservation purposes started in the mid 19th century. Countries have also enacted progressively stronger laws for the protection of individual species.

1864 Yosemite Grant Act passed by US President Abraham Lincoln establishing first major modern protected area

1872 Yellowstone National Park, USA, the world's first national park, established

1948 The IUCN, then called the International Union for the Protection of Nature (IUPN), founded

1958 The IUCN establishes provisional National Parks Commission

17% of Earth's land surface is in some form of nature reserve or national park

The regional picture

All regions of the world have designated protected areas, but many are not properly implemented. Scientists have determined that it would cost about 0.12 per cent of global GDP to rectify this, as well as to enforce other conservation measures. Meanwhile, the global cost of environmental damage is estimated at about 11 per cent of global GDP.

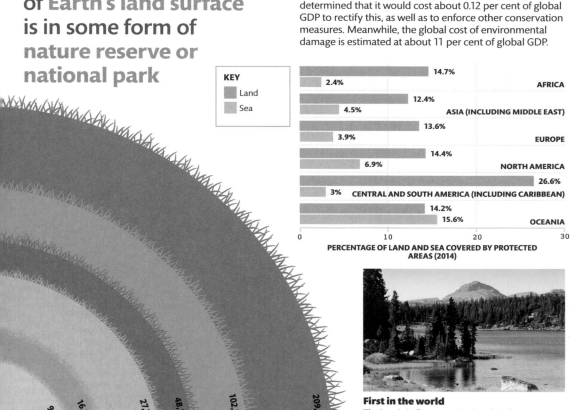

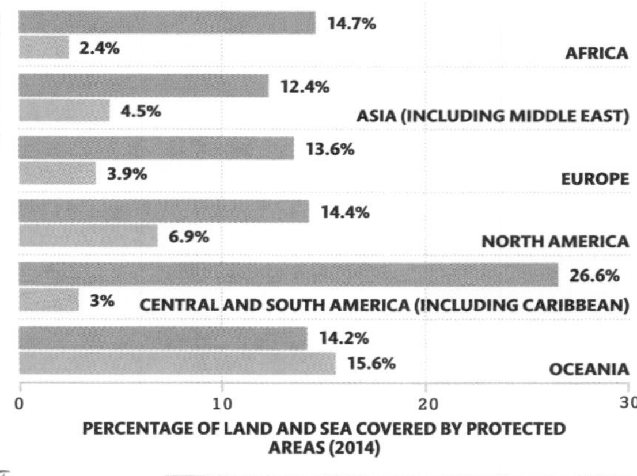

PERCENTAGE OF LAND AND SEA COVERED BY PROTECTED AREAS (2014)

9,214 SITES

16,394 SITES

27,794 SITES

48,388 SITES

102,102 SITES

209,429 SITES

First in the world

The iconic Yellowstone National Park was established in 1872. Today it protects one of Earth's last remaining nearly intact temperate zone ecosystems.

1962 First World Parks Congress, a global forum on protected areas, held in Seattle, USA

1972 United Nations Environment Programme and World Heritage Convention established

1982 Third World Parks Congress focuses on protected areas and sustainable development

1992 UN Convention on Biological Diversity (CBD) treaty agreed at Rio de Janeiro Earth Summit

2010 CBD adopts Aichi Biodiversity Targets to halt loss of biodiversity

2015 UN Sustainable Development Goals adopted (see pp184–85), including targets for the protection of nature

Shaping the future

Since the start of the first industrial revolution, successive waves of invention have driven economic development and led to improved living conditions for billions of people. Many factors have shaped innovation. These include access to natural resources, the strength of the societies that develop new technologies, the role of government in encouraging innovation, levels of education, and how existing technologies provide springboards for new invention. A new wave of innovation is breaking and could be vital in enabling development that respects the planet.

Waves of innovation

Since the middle of the 18th century, there have been a number of new industrial revolutions. Each of these has reshaped every aspect of the economy and society, and they have all followed a similar pattern, with the initial invention creating a period of boom and rising wealth. In the process, this gave rise to secondary economies based on core inputs, such as coal for steam engines and computer chips for the computers that drive the digital economy. Each time the technology reaches maturity, it is subject to a period of adjustment before, ultimately, being replaced. History reveals successive waves of progress driven by new technologies that last for about 50 years each. We could be at the start of a new one – the sustainability revolution.

Second wave: Steam power
Water is superseded by coal-fired steam engines. They drive manufacturing and long-distance transport by rail and ships. Global trade rapidly expands.

First wave of innovation: Water power
Water-powered machinery – driven by mills on streams – transform textile manufacture, and leads to the industrialization of work previously done by individual workers.

1785 1800 1820 1840 1860 1880

YEAR

BIOMIMICRY

Biomimicry is the process of mimicking nature. For example, termites cool their mounds by using vents to circulate air. Architects created the Eastgate Centre in Zimbabwe with an air conditioning system that is modelled on these termite mounds. It uses minimal electricity, which has resulted in drastically reduced carbon emissions.

90% **The energy savings from using biomimicry** for ventilation in the Eastgate Centre, Zimbabwe

Sixth wave: Sustainability
A new industrial revolution is built on sustainability. This uses renewable energy, the restoration of ecosystems (to provide essential services), zero-waste circular economy products, sustainable farming, biomimicry, and innovations in nanotechnology.

Third wave: Electrification
Electrical power transforms the world, along with the rise of the internal combustion engine, which revolutionizes transport with fossil oil.

Fourth wave: Space age
Aviation technologies are refined to provide long-distance mass transport and take us into space. Electronics and petrochemicals transform the lives of consumers.

Fifth wave: Digital world
Computers go mainstream, changing our lives, as well as business and government. Biotechnology and other industries develop as the digital revolution gathers pace.

Low carbon growth

The sustainability revolution will need to meet the demands of more people while achieving a drastic reduction in environmental impact. Carbon emissions provide an illustrative case in point.

Present patterns of economic development are carbon intensive. In other words, for every unit of economic output, we are producing high levels of carbon dioxide (CO_2). The strategy going forwards needs to be towards a less carbon-intensive society – a world where we can continue to grow in wealth but are less dependent on factors (like fossil fuel energy production) that increase CO_2 emissions. This "delinking" of economic growth from carbon emmisions is vital if we are to have any reasonable chance of ensuring the average global temperature increase does not exceed two degrees.

World carbon intensity

768 gram CO_2/ US$

Japan

244 gram CO_2/ US$

United Kingdom

347 gram CO_2/ US$

Carbon intensity

This graphic shows the carbon intensity of each dollar of GDP in 2007 in the UK and Japan, as well as the world's average levels. The relatively efficient use of energy, gas generation, and some nuclear power leads the UK to emit at about half the global GDP average. Lacking fossil energy resources, Japan's economy is quite efficient. The country uses a great deal of nuclear power and generally has a low level of emissions per unit of GDP compared with the world average. However, both countries are far off the much lower global carbon target needed by 2050 – from six to 36 grams CO_2 per dollar (see Scenarios 1–4, oppposite).

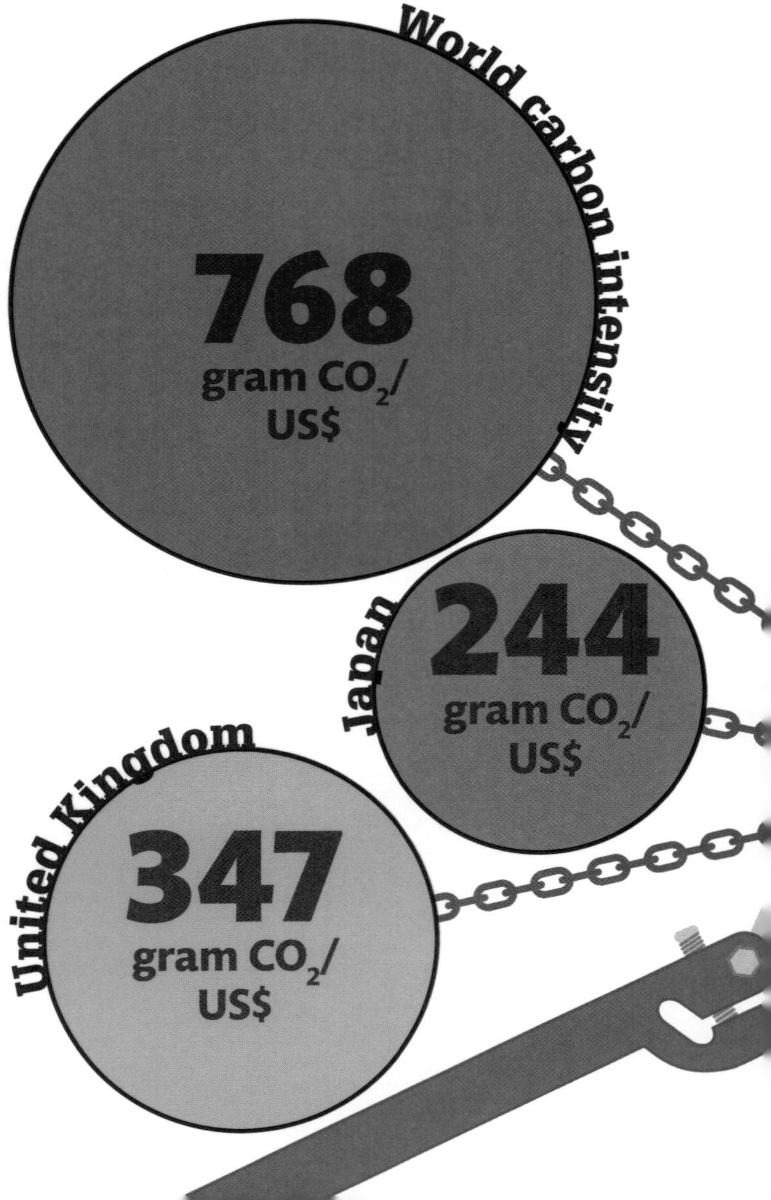

Future scenarios

In 2007, the economist Tim Jackson set out the scale of emissions reductions needed to avoid a 2°C (3.6°F) global temperature increase, compared with preindustrial times. To reveal the scale of action needed, he came up with four scenarios. Each scenario uses variations on population number, the level of inequality, and average income to determine how much carbon emissions must be reduced compared with 2007. If the world is to achieve outcomes in scenario 4, then the carbon intensity of each dollar of GDP must drop to six grams, or by about 6 per cent annually. However, compared with 2007, in 2019 global carbon emissions had increased by about 16 per cent.

6.2%

The amount that the **global economy** needs to cut carbon intensity **each year**

GDP

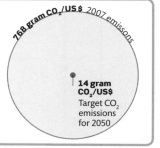

2050 Scenario 1

Assumes that population grows to nine billion. Per capita income growth continues at the 2007 level, but inequalities remain.

WORLD POPULATION 9 BILLION

PER CAPITA INCOME GROWTH $ ↑

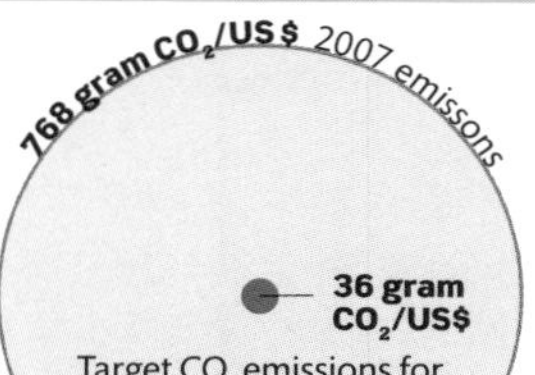

2050 Scenario 2

Assumes population grows to 11 billion. As in scenario 1, per capita income growth continues at the 2007 level, but inequalities remain.

WORLD POPULATION 11 BILLION

PER CAPITA INCOME GROWTH $ ↑

768 gram CO_2/US$ 2007 emissons

30 gram CO_2/US$ Target CO_2 emissions for 2050

2050 Scenario 3

Population grows to nine billion (as in scenario 1). Everyone enjoys per capita income at the equivalent of the EU average in 2007.

WORLD POPULATION 9 BILLION

PER CAPITA INCOME GROWTH $ ↑↑

768 gram CO_2/US$ 2007 emissons

14 gram CO_2/US$ Target CO_2 emissions for 2050

2050 Scenario 4

Population grows to nine billion. Everyone enjoys high living standards because of economic growth above those of the EU today.

WORLD POPULATION 9 BILLION

PER CAPITA INCOME GROWTH $ ↑↑↑

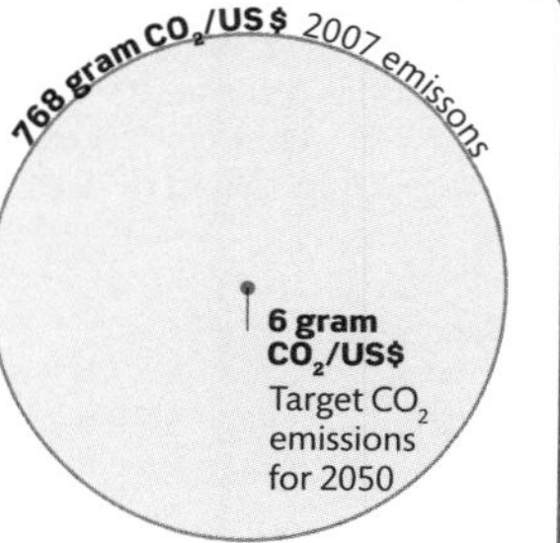

The rise of clean technology

Increasing our use of clean technology through harnessing renewable energy, promoting energy efficiency, recycling, green transportation, and the more rational use of water is vital to reduce our ecological footprint.

Clean technology is beginning to make a difference. Most notably, switching to renewable energy sources decreases the amount of carbon that would have been released through burning fossil fuels. Other promising developments include technology that extracts resources from what would otherwise be waste, more efficient water treatment, nutrient recovery facilities that prevent pollution, and information technologies that enable buildings to run more efficiently. Clean technology companies are attracting increased investment as they become more efficient and competitive, which is helping them to grow. From 2007–2010 the clean technology sector expanded by, on average, 11.8 per cent per year and in 2011–2012 it comprised a market worth around US$5.5 trillion.

Developing a clean future

Clean technology is driving growth in developing countries, including among small- and medium-sized enterprises (SMEs). A World Bank study estimated that from 2014–2024, US$6.4 trillion will be invested in clean technology in developing countries, with US$1.6 trillion of that accessible to SMEs. South America and Sub-Saharan Africa are predicted to be major growth areas in developing-world clean technology.

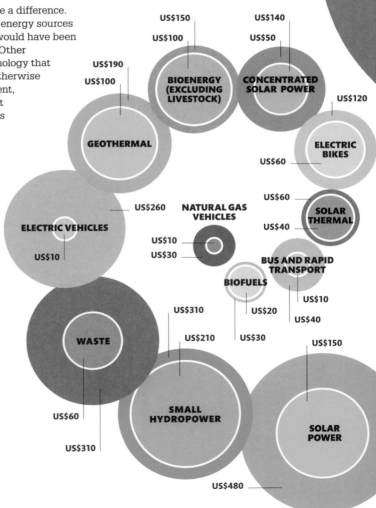

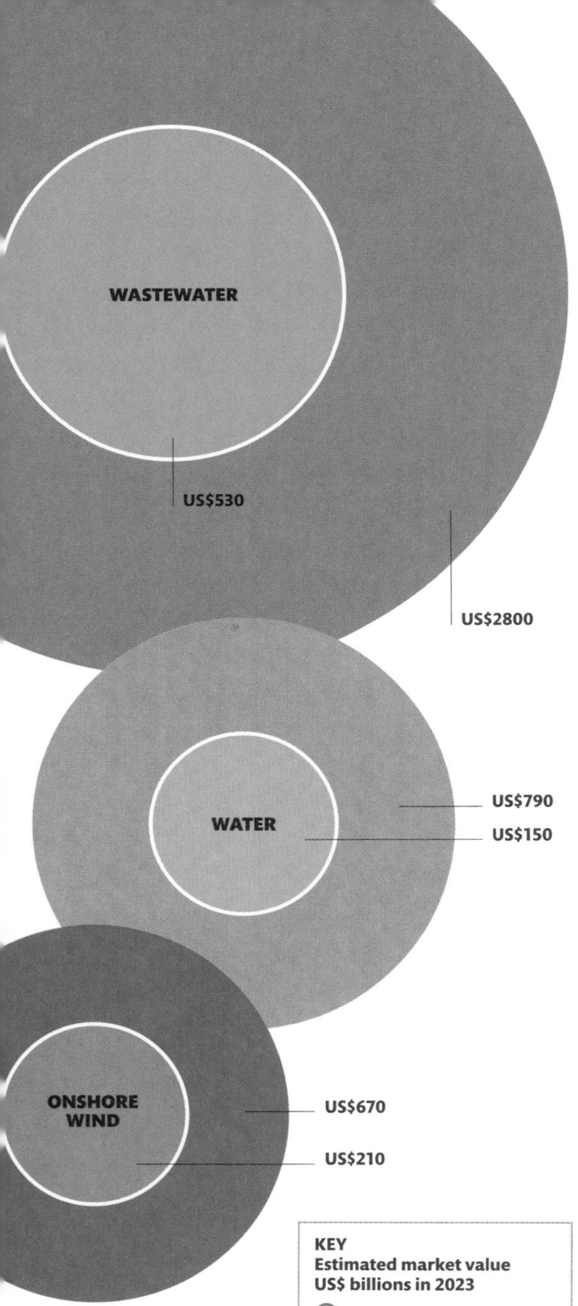

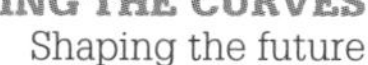

WASTEWATER
US$530
US$2800

WATER
US$790
US$150

ONSHORE WIND
US$670
US$210

KEY
Estimated market value US$ billions in 2023
Total
SME share

Clean, green jobs

According to the International Renewable Energy Agency, there were 11.5 million jobs in renewable energy in 2019, up from 7.2 million in 2012. The largest concentrations are in China, Brazil, the USA, India, and Germany. Solar power, hydroelectric power, and biofuels are the biggest employers.

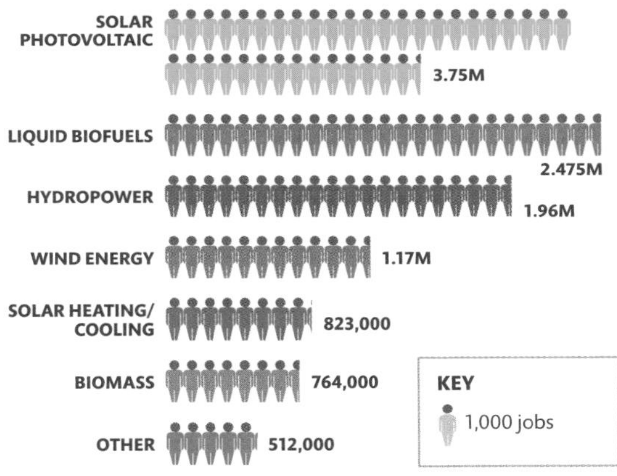

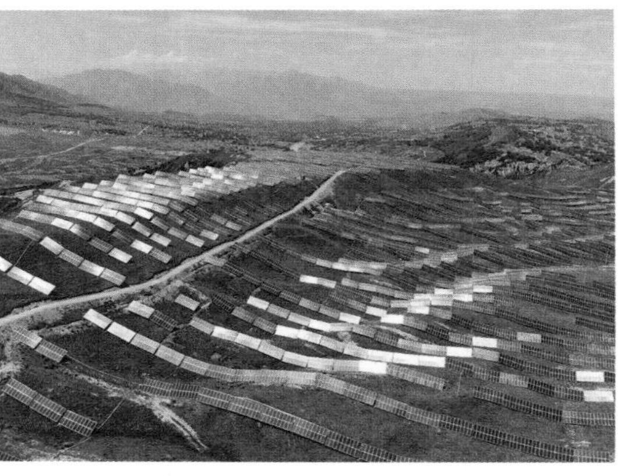

Energy farm
The deployment of solar photovoltaic technologies is rising fast in Asia, as in this installation in China. In 2019, Asia provided 63 per cent of total jobs in renewables globally.

A sustainable economy

If the world is to achieve the Sustainable Development Goals (see pp184–85) to raise living standards while avoiding the worst impacts of climate change, resource depletion, and ecosystem degradation, then economic change is needed.

Rewiring the economy

In 2015, the University of Cambridge Institute for Sustainability Leadership (CISL) in the UK proposed a plan to "rewire" the economy. The plan sets out 10 tasks for governments, business, and financial institutions to place our economic system more in line with social and environmental priorities. The tasks (right) relate to changes in government policy and the world of business while harnessing the massive power of finance. The changes are deliberately geared to promoting the achievement of the Sustainable Development Goals, which cannot be reached through traditional environmental and development programmes on their own. A more fundamental shift is needed. That shift goes to the heart of our economy.

"If there is waste or pollution, **someone along the line pays for it.**"

LEE SCOTT, FORMER CHIEF EXECUTIVE, WALMART

Government

- **Set the right targets and measures**
  For example, official goals to cut greenhouse gas emissions and protect ecosystems must be backed by policies to meet them.
- **Introduce new tax systems**
  Show the true cost of different choices – such as taxing waste and pollution to promote cleaner production and energy sources.
- **Positive influence**
  Drive positive change by harnessing the power of public spending, subsidies, planning rules, education, and research.

Finance

- **Ensure that capital acts for the long term**
  Extend the timeframes over which financial risks and returns are modelled, thereby reducing short-term decisions while protecting investors.
- **Value the true costs of business activity**
  Identify strategies that encourage companies to meet social and environmental goals while they pursue financial profitability.
- **Innovate financial structures**
  Make finance work for social benefit, including fighting climate change and protecting the planet's ecosystems.

Business

- **Set bold ambitions**
  Transform company activities to embrace goals for low-carbon energy, zero deforestation, and zero waste.
- **Broaden measurement and disclosure**
  Ensure that companies report on the full range of impacts they create, including social and environmental performance.
- **Grow capability and incentive**
  Harness companies' talent and money by, for example, linking executive bonuses with reduced carbon emissions.
- **Harness the power of communications**
  Change advertising to avoid messages that undermine social and environmental progress.

Despite the many social and environmental pressures facing our world today, the achievement of the Sustainable Development Goals could lay the foundations for a very positive future. This requires a change in mind-set, however, and moving beyond the view that environmental protection brings unaffordable financial costs. In reality, social progress cannot be achieved if the natural environment continues to degrade. That is why environmental damage must be minimized through the way the economy operates. There is evidence from around the world that this is beginning to happen, as policies, investment patterns, and business practices begin to change.

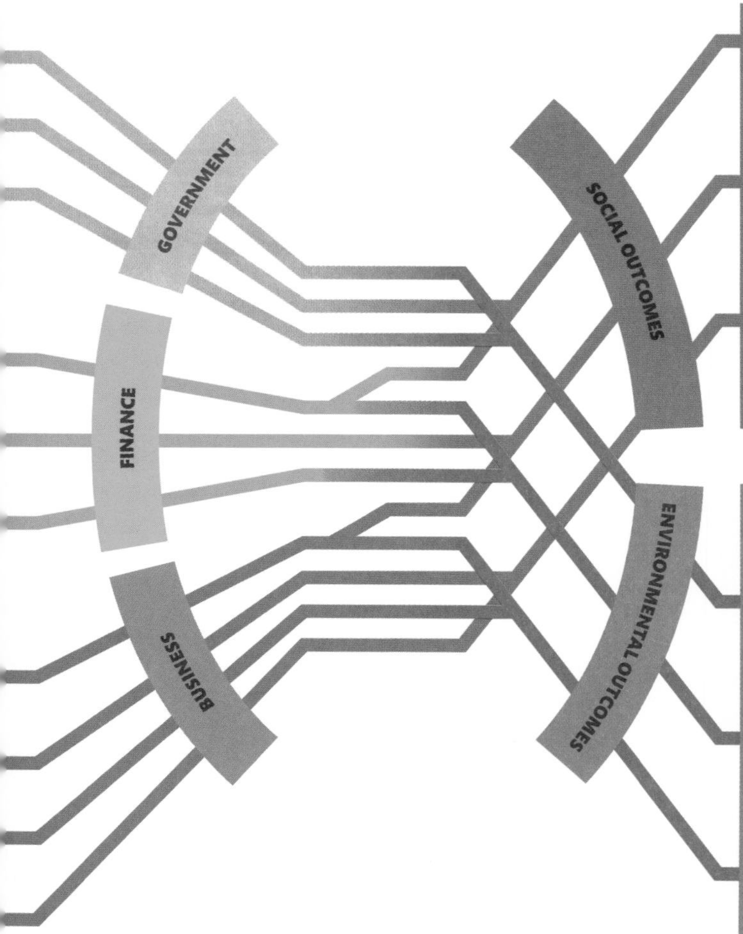

Social outcomes

› Basic needs
Lead to stronger drivers for meeting basic needs, including food, water, shelter, sanitation, security, and basic freedoms.

› Health and well-being
Investment in health and education enhances wellbeing, and contributes to economic development and environmental goals.

› Decent work
Economic strategies generate many more socially inclusive jobs with good working conditions.

Environmental outcomes

› Preserving resources
Economic plans lead to the cleaner and more efficient use of natural resources and the development of a "circular economy" (see pp200–01).

› Ecological maintenance
Economic planning leads to the promotion of ecosystem health, not least because this is essential for future economic development.

› Temperature target
Economic development strategies embrace low-carbon outcomes, stabilizing greenhouse gases in the atmosphere so as to limit average global temperature increase to well below two degrees Celcius.

Circular economy

Centuries of development and economic growth have been founded on a largely linear economy. This system takes resources – such as fossil fuels, metals, and nutrients – uses them, and then disposes of waste to air, water, and land. While this has sustained population growth and achieved more comfortable living standards, it has had many negative consequences, including climate change, resource depletion, pollution, and ecosystem damage. A circular economy, by contrast, reduces these impacts by treating waste as new resources. There are two illustrative examples of how a circular economy works – one biological and one material. The same basic ideas can be applied across the economy using a variety of biological nutrients and materials.

Sewage treatment plant
New technology is already fitted to some sewage works. Phosphorus is captured from waste and turned into a high-quality fertilizer.

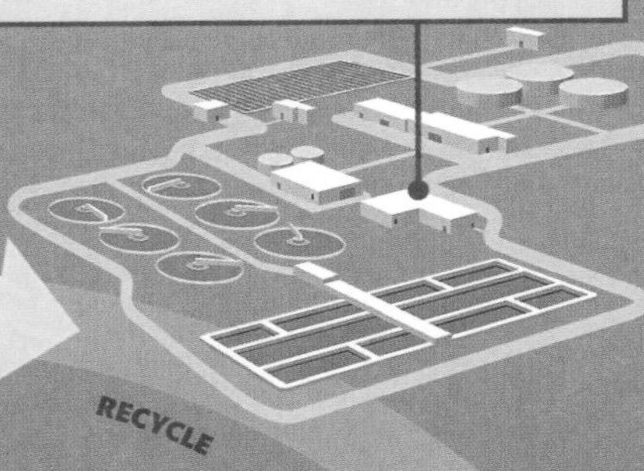

RECYCLE

Biological cycle

Phosphorus is an essential biological nutrient. In our linear economy, we mine phosphorus from finite rock sources. It is then dispersed in the environment, causing ecosystem damage. In a circular economy, phosphorus is recycled to sustain new plant growth. This saves resources and protects the environment.

CONSUME

Consumption
Food that is eaten passes through the human digestive system. Waste is transmitted to sewage treatment works via toilets.

Starting point
Biological materials, such as phosphates, originally come from nature. If these are reused it limits the need to extract more.

GROW/USE

Food supply and sale
Food is supplied to shops, supermarkets, and markets. Part of the cost of the produce is determined by the price of fertilizer, such as phosphorus.

Growing crops
Phosphorus is applied to fields as fertilizer to promote plant growth and increase the crop yields needed to feed a growing population.

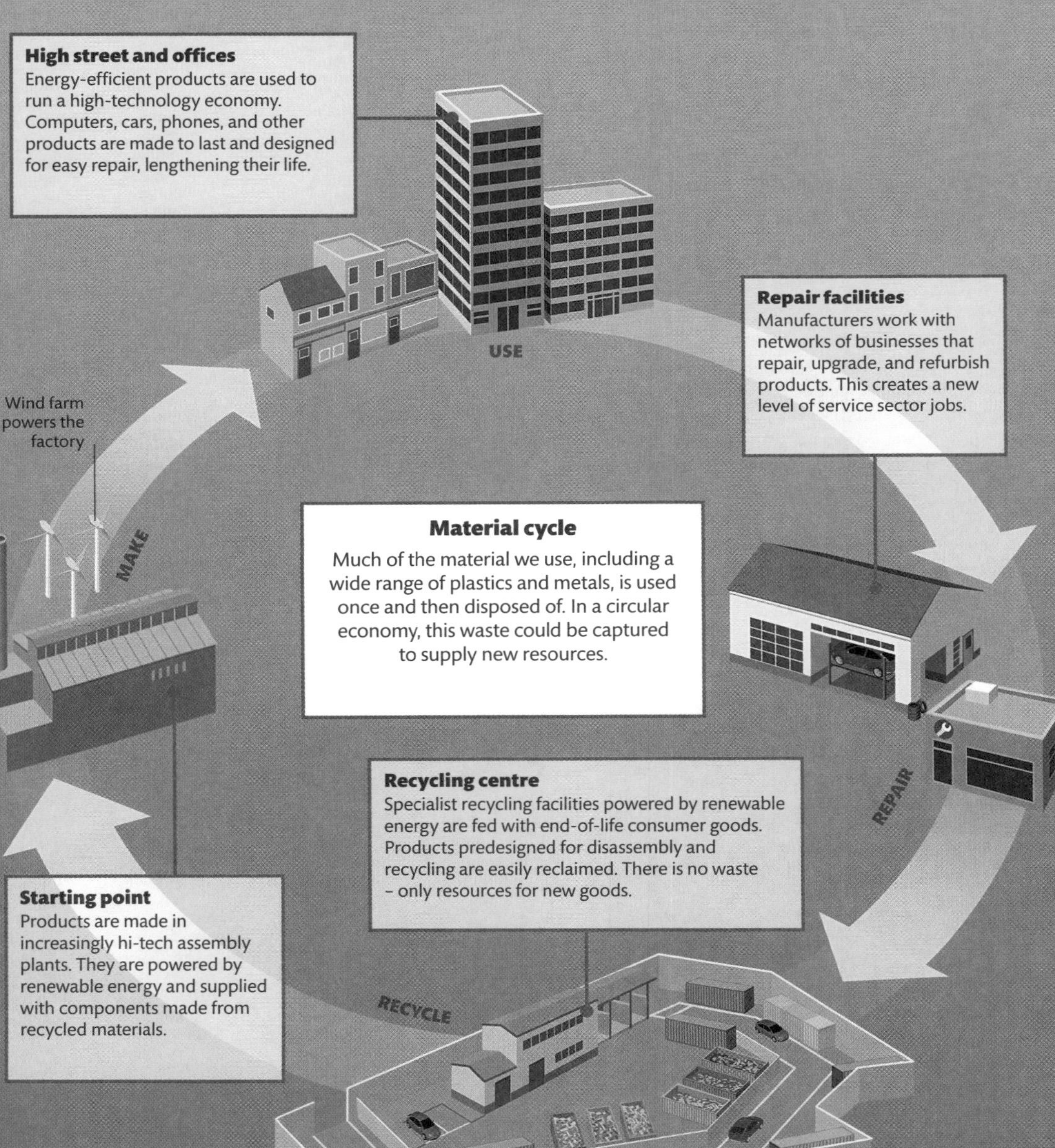
High street and offices
Energy-efficient products are used to run a high-technology economy. Computers, cars, phones, and other products are made to last and designed for easy repair, lengthening their life.
USE
Repair facilities
Manufacturers work with networks of businesses that repair, upgrade, and refurbish products. This creates a new level of service sector jobs.
Wind farm powers the factory
MAKE
Material cycle
Much of the material we use, including a wide range of plastics and metals, is used once and then disposed of. In a circular economy, this waste could be captured to supply new resources.
REPAIR
Recycling centre
Specialist recycling facilities powered by renewable energy are fed with end-of-life consumer goods. Products predesigned for disassembly and recycling are easily reclaimed. There is no waste – only resources for new goods.
Starting point
Products are made in increasingly hi-tech assembly plants. They are powered by renewable energy and supplied with components made from recycled materials.
RECYCLE

A new mindset

Taking resources from nature and releasing wastes into the biosphere is causing mounting environmental pressures that threaten progress. New patterns of development are needed.

Our growing demands upon nature have caused profound changes to the systems that sustain life on Earth. These changes are now having huge economic and humanitarian impacts. A shift in approach is needed, so that rising human demand is no longer met at the expense of the environment, but instead embraces the restoration and protection of ecological systems. This in turn leads to the need for an approach that achieves sustainable economic development and improved social conditions while respecting ecological limits.

The safe zone

UK economist and sustainable development expert Kate Raworth proposes the idea of "doughnut economics", whereby social and ecological factors are equally respected. At the moment, one (social progress, such as better health, jobs, and education) is built on the sacrifice of the other (ecological systems). This graphic demonstrates the doughnut concept. The outer ring is the environmental "ceiling", made up of nine planetary boundaries (see pp178–79). Beyond these limits are unacceptable levels of environmental damage. The inner ring consists of 10 social factors, below which are unacceptable levels of human deprivation. Between the two rings is a doughnut-shaped space which is both environmentally safe and socially just: the space where all of humanity can thrive.

ENVIRONMENTAL "CEILING"

FRESHWATER USE
Damage to ecosystems and wasteful water use threaten to increase water stress and undermine food security.

CLIMATE CHANGE
Global warming will increase risks of food shortages, water stress, conflict, and spread of disease.

SUSTAINABLE ECONOMIC DEVELOPMENT

WATER

FOOD

LAND USE CHANGE
As more land is taken over for farming and urbanization, a series of essential ecosystems are being degraded.

HEALTH

SOCIAL EQUALITY

ENERGY

BIODIVERSITY LOSS
All our food and many medicines are ultimately derived from wild species. Biodiversity is vital for a sustainable future.

OZONE DEPLETION
This is a serious threat to human welfare as rising levels of ultraviolet radiation increase the risk of skin cancer.

NITROGEN AND PHOSPHORUS
Rising levels of these nutrients in the environment are damaging fish stocks (pp154–55) and threatening human health.

OCEAN ACIDIFICATION
Marine plankton species that replenish atmospheric oxygen could be threatened by ocean acidification (pp152–53), caused by increased carbon dioxide.

CHEMICAL POLLUTION
Toxic materials affect natural diversity, including beneficial wildlife, such as pollinators that underpin a secure food supply (pp68–69).

ATMOSPHERIC POLLUTION
Dust, smoke, and haze have increased in the air because of human activities, and these pose a serious threat to people's health.

COME

EDUCATION

HAVING A VOICE

RESILIENCE

KEY TO THE RINGS
- Environmental ceiling
- Sustainable economic development
- Social foundation

PLANETARY STRESS

Oxfam estimates that one tenth of the population is most responsible for factors that result in planetary stress, such as greenhouse gas emissions and energy use. It is their consumption, and the production methods of the companies producing the goods and services these wealthiest people buy, that drives most of the environmental damage threatening human security.

EMISSIONS
Half the world's carbon dioxide emissions are generated by 11 per cent of the world's population.

ENERGY
High-income countries are home to 16 per cent of the world's population, but use 57 per cent of all electricity.

PURCHASING POWER
The same 16 per cent of the global population also account for 64 per cent of all spending on consumer goods.

NITROGEN (FOOD)
The EU has 7 per cent of the world's population, but uses 33 per cent of the planet's sustainable nitrogen budget to grow and import animal feed.

What can we do?

- **Governments** across the world must adopt the 2015 Sustainable Development Goals at the heart of their economic strategies.
- **Companies** must align their plans with long-term sustainability, protecting social and ecological values.

What can I do?

- **Elect politicians** who are supporters of "doughnut economics".
- **Buy from firms** building "doughnut economics" into their business strategies.
- **Support campaigns** promoting human welfare within planetary limits.

Restoring the future

If we are to lay the foundations for a secure future, then the centuries of environmental degradation must be halted and reversed. This is both an economically rational and achievable priority.

Our approach to development and economic growth so far has assumed that the sacrifice of environmental ecosystems and the pollution of air and water are inevitable prices of progress. While this growth has brought comfort, convenience, and security to billions of people around the world, we are in a period of diminishing returns. The damage caused by climate change, air and water pollution, depletion of resources, and ecosystem deterioration threatens to exceed all the benefits of growth. But it is still possible to restore environmental health through sustainable development.

Restoration in progress

Continuing environmental degradation is not inevitable, and it can be reversed if we decide to build on positive examples from around the world that already demonstrate what is possible. From Brazil to Denmark, and from Uruguay to Bhutan, there are hundreds of inspirational examples of what can be done across a range of sectors, including farming, transport, conservation, infrastructure, and energy supply. Governments, international agencies, businesses, and individual citizens all need to play their part in the sustainability transformation necessary in the 21st century.

Natural environment

The protection of nature is a sound economic investment, but failing to see this is causing ecosystem degradation and mass extinction of animals and plants.

Agriculture

Climate change, water scarcity, damaged soils, and the decline of beneficial animals, such as bees, are all major threats to future food security.

Infrastructure

Current approaches towards the expansion and development of built-up spaces "lock in" wasteful, high-carbon, and resource-intensive patterns of living.

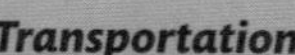

Transportation

Air pollution, congestion, and climate change are among the expensive consequences of our transport system. Commuting wastes time and causes stress, while clogging up city streets.

Energy supplies

High-carbon emissions and dangerous air pollution cause widespread damage. Wasteful energy consumption increases the negative impact on the environment.

Present

Natural environment

The realization that thriving nature is essential for healthy people, strong society, and a sound economy brings an end to environmental damage and leads to restored ecosystems.

Agriculture

Sustainable farming that protects soil, water, and wildlife, and changes in the wider food system, including a huge reduction in food waste, result in secure nutrition, which causes less environmental damage.

Infrastructure

The cities of the future are designed to be efficient and pleasant places to live. Engineering and ecology combine to create truly sustainable, healthy cities.

Transportation

Cycling and walking improve public health, cut pollution, and reduce emissions. Digital technologies allow home-working, which reduces commuting. Electric vehicles mean cleaner transport.

Energy supplies

Greenhouse gas emissions are reduced by the efficient use of renewably generated heat and power. Clean electricity charges the batteries of electric vehicles.

Future

"Making **peace with nature** is the **defining task** of the 21st century."

ANTÓNIO GUTERRES, UNITED NATIONS SECRETARY-GENERAL

What can we do?

- **Investors can adopt strategies** that direct finance towards positive solutions, such as renewable energy and sustainable farming.
- **Governments can introduce incentives** to adopt clean technology, including subsidies to encourage the protection and restoration of ecosystems.

What can I do?

- **Choose products and services** from companies offering solutions for sustainability challenges. This rewards market leaders and puts pressure on those companies lagging behind.
- **Urge your bank** and pension fund to lend and invest only in enterprises that back a secure and sustainable future.

Glossary

UNITS

MTOE - million tons of oil equivalent The amount of energy released by burning one million tonnes of oil; used as a measure of energy production or consumption.

MWh - megawatt hour A measure of electrical energy use. 1 megawatt is 1 million watts; 1 MWh is the power of 1 million watts used or produced constantly for 1 hour.

TWh - terawatt hour A measure of electrical energy use. 1 terawatt is 1 trillion watts; 1 TWh is the power of 1 trillion watts used or produced continuously for 1 hour. See also MWh.

BTU - British thermal unit The amount of heat required to raise the temperature of 1 lb of water by 1°F at sea level. Used to measure the heat output of heating and cooling systems and machines.

DU- Dobson unit A unit used for measuring the concentration of trace gases, notably ozone, in the atmosphere.

gigatonnes (billions of tonnes) of CO_2 A unit of measurement used for carbon dioxide or carbon emissions. One gigatonne is one billion tonnes. One tonne of carbon is equivalent to 3.67 tonnes of carbon dioxide. The difference arises from the fact that $GtCO_2$ includes the weight of two units of oxygen. Used to measure carbon emissions. A similar unit, $GtCO_2$-eq (gigatonnes of CO_2 equivalent), may be used to measure other greenhouse gases in the "common currency" of the warming caused by carbon dioxide. To convert carbon dioxide to carbon, divide by 3.67. For example, 1Gt of CO_2 is equivalent to 272 million tonnes of carbon.

Ej - exajoule Unit of energy equivalent to 1 billion gigajoules. (A gigajoule is 1 billion joules.)

GtC/GtC02 -Gigatonnes of carbon/ Gigatonnes of carbon dioxide. One gigatonne is one billion tonnes. One tonne of carbon is equivalent to 3.67 tonnes of carbon dioxide. The difference arises from the fact that GtCO2 includes the weight of two units of oxygen. Used to measure carbon emissions.

ng (nanogram) One billionth of a gram.

GENERAL

acid rain Rain, sleet, or snow contaminated with air pollutants such as sulphur dioxide and nitrogen oxides. Acid rain pollutes soil and water and damages buildings.

acidification A process by which oceans, lakes, or rivers gradually become more acid. Acidification of oceans is largely due to increased uptake of CO_2 from the air. In lakes and rivers, it may be due to acid rain entering the water.

algal bloom Rapid growth of algae in a lake or ocean, often due to an excess of nutrients such as nitrogen or phosphorus. Algae can block out sunlight and use up oxygen. Some algal blooms produce toxins harmful to animals or humans.

atmosphere The layer of gases surrounding the Earth (or any other planet). Earth's atmosphere consists mainly of nitrogen (78%) and oxygen (21%).

biodegradable A term used for materials that can be broken down naturally by microorganisms into constituent molecules or elements.

biodiversity Variety in living things. Species biodiversity is the variety of species in an environment. Genetic biodiversity is the variation in genes within one species. Ecological diversity is the range of ecosystems and habitats.

bioenergy Renewable energy extracted from biological materials such as wood, straw, manure, and sewage.

biofuels Generally used to describe liquid fuels derived from plants and other organic material, such as food waste, providing alternatives to petrol, diesel, and kerosene. Biogas is an alternative to fossil gas, also made from organic material, such as animal waste or food waste.

biogeochemical flow The circulation of a chemical substance, such as carbon or nitrogen, through the atmosphere, soil, biosphere (plants and animals), and water.

biomagnification The process by which a chemical (such as a pesticide) becomes more concentrated as it passes through a food web, for example as filter-feeding organisms are eaten by bigger creatures and finally top carnivores.

biomass The mass of living organisms (plants, animals, microorganisms) in a given ecosystem or community.

biome An area of land surface, freshwater, or ocean characterized by particular types of vegetation, as well as by physical features such as climate or water depth.

biomimicry Imitation of natural structures and processes to help meet challenges in the human world.

bioproductivity The rate of production of biomass from a specific ecosystem over a given time period.

biosphere The zone of the Earth containing all living organisms; comprises the Earth's surface, oceans, and the lowest layer of the atmosphere.

carbon A common chemical element (symbol: C) that binds with other elements such as hydrogen (H) and oxygen (O) to form compounds such as carbon dioxide. Carbon is found in all living organisms.

carbon capture and storage (CCS) A process in which carbon dioxide from burning fossil fuels is captured before it reaches the atmosphere, then deposited deep inside rocks.

carbon dioxide A gas with molecules formed from one carbon atom and two oxygen atoms (formula: CO_2); produced by respiration from living organisms, fermentation of dead matter, and combustion (fires, or burning of biofuels or fossil fuels).

carbon intensity A measure of greenhouse gas emissions calculated as the mass of carbon emitted per unit of energy consumed. . Carbon intensity can also be calculated in relation to emissions per unit of GDP. In this case, the concept can also embrace emissions from deforestation as well as energy.

carbon pricing A tax or market price levied on emissions of carbon dioxide to incentivize changed behaviour, such as more efficient energy use or the expansion of renewable energy.

carbon sink An ecological system that absorbs and stores carbon dioxide from the atmosphere. Oceans and forests are the Earth's two main carbon sinks.

carrying capacity The maximum population size of a species that an ecosystem or habitat can support indefinitely.

chlorofluorocarbons (CFCs) Chemical compounds formed from chlorine, fluorine, and carbon. CFCs were widely used in refrigeration, as propellants for aerosols, and as solvents, but it was found that they damaged the ozone layer so their use is now restricted.

climate The average atmospheric conditions in an area over a long period of time. It is influenced by the latitude and elevation of an area, plus factors such as average temperatures and rainfall.

CO_2 emissions Release of carbon dioxide by natural means (as in forest fires and volcanic eruptions) or artificial means (such as burning of fossil fuels).

consumption (economic) The purchase and use of goods and services by individuals or households.

convection The transfer of heat through movement of a fluid (such as air or water). For example, in convection cells in the atmosphere (see pp128–29), warmer air expands and rises, while cooler air sinks, creating air currents.

dead zone An area of a lake or ocean where the water is so low in oxygen that many animals cannot survive there. Dead zones can result from *algal blooms* caused by water pollution.

deforestation Destruction and/or removal of trees from an area of forest, to leave open land. Major causes include logging or clearance of forest for ranches or plantations of crops. Deforestation can lead to soil erosion and loss of *biodiversity*.

desalination Removal of salt and other minerals from water to make the water suitable for drinking or irrigation.

desertification The spread of desert conditions to areas that were previously covered with vegetation; caused by factors that include reduced rainfall and over-grazing by domesticated animals

developed country A country with a relatively stable industrial or post-industrial economy, established political security, advanced level of technology, and generally high standard of living compared with other nations.

developing country A country with a weak infrastructure, insufficient public services, and in which the majority of people have relatively low income, lower life expectancy, and limited access to comprehensive modern health care and education.

dieback In trees and shrubs, the progressive death of twigs, then branches, then the whole plant. Possible causes include infection, pest infestation, drought, and pollution.

dioxin A group of persistent chemicals that may be emitted via industries such as paper bleaching and processes such as waste incineration. These chemicals are toxic and can pose risks to animal and human health through bioaccumulation in food chains.

DUs – Dobson units See *Units*

E7 countries A group of seven powerful countries with emerging market economies: China, India, Brazil, Russia, Mexico, Turkey, and Indonesia. The E7 now accounts for around 30 per cent of world *GDP*.

ecology The science that deals with inter-relationships between organisms and each other and their non-living environment, including air, water, and geology.

ecosystem A self-sustaining community of living things interacting with each other and with their natural environment.

El Niño A large-scale climate disturbance occurring about every 3–7 years in the central and eastern equatorial Pacific Ocean, in which warming of ocean surface currents gives rise to changes in prevailing weather patterns across the world, but especially along the coasts of North and South America and north of Australia. See also *La Niña*.

emerging market A national economy that is growing, developing, and industrializing rapidly from a relatively low income and economic base compared with the already developed nations. Many of these countries are becoming increasingly powerful in industry, trade, and technology.

emissions Discharges of gases, liquid vapours, and tiny particles into the atmosphere; usually refers to discharges from human sources such as vehicles, power stations, and deforestation.

energy storage The collection and storage of electrical or mechanical energy for use at a later time, either on a small scale (as in a rechargeable battery) or a large scale (as with a reservoir for a hydroelectric plant).

erosion A process in which soil or rock is broken down and carried away by wind, flowing water, or ice. Erosion processes may be mechanical (in which rock or soil is physically worn away) or chemical (in which the rock or soil is dissolved in water).

eutrophication Ecological change resulting from the rising concentration of nutrients, such as nitrates and phosphates, in an ecosystem such as a body of water. Eutrophication can give rise to *algal blooms* and *dead zones*.

evaporation The process in which molecules from the surface of a liquid change to a vapour, usually due to increased temperature, as when water evaporates from a sea or lake on a warm day.

extinction The disappearance of a species, subspecies, or group of organisms marked by the the death of the last individual.

flood plain A flat area of land beside a river that naturally floods whenever the water level rises above the level of the river banks.

food chain or food web A hierarchy (chain) or network (web) of organisms in which those at one level eat others; for example, a bird of prey species may eat insect-eating birds, which in turn eat insects, which in turn eat plants.

food miles/food kilometres The distance that food has travelled from the place where it was produced to reach the consumers. Longer distances involve higher fuel use, so cutting food miles/km can help reduce emissions from transport.

food security The state that exists when people have access to, and can afford, enough nutritious food to maintain a healthy life.

fossil fuel A fuel produced from the remains of plants and animals that died tens or hundreds of millions of years ago, such as coal, oil, or *natural gas*. These fuels contain *carbon* captured from the *atmosphere*, so when burned release *carbon dioxide* into the atmosphere.

fracking (hydraulic fracturing) Injection of a high-pressure mix of water, sand, and chemicals into oil- or gas-bearing rock to create cracks (fracturing) and release the oil or gas. Fracking can lead to contamination of groundwater and may even trigger small earthquakes.

G7 countries A group of seven major industrialized countries – the USA, Canada, UK, France, Germany, Italy, and Japan – whose leaders and finance ministers meet annually to discuss global economic policy and international security.

GDP per capita A measure of economic performance; it is calculated by dividing a country's GDP by the number of people in the population.

GDP – Gross Domestic Product The monetary value of all the finished goods and services produced in a country over a specific time (usually a year). See also *Real GDP*.

geothermal energy Energy derived from heat generated naturally within the Earth, for example obtained from hot springs in areas of volcanic activity.

global warming The increase in the average temperature of the *atmosphere* and/or oceans, that in turn affects the extent of ice on earth, sea levels, and *weather*, including rainfall. Human activities have recently played a fundamental role in raising global temperatures.

green revolution A set of advances in crop cultivation, starting in the 1940s, that vastly increased food supplies, especially in developing countries.

greenhouse effect The process by which the Earth's atmosphere traps more energy from the Sun, thus warming the atmosphere and oceans.

greenhouse gas A gas that traps heat in the *atmosphere*. The main gas is *carbon dioxide*; other major gases are *methane* and *nitrous oxide*. Greenhouse gas emissions from human activities such as burning fuels contribute to *global warming*.

groundwater Water held within the spaces in soil and rock, notably in water-saturated rocks called aquifers.

gyre A large system of ocean currents rotating in a spiral.

Haber-Bosch process A synthetic process in which nitrogen from the air is combined with hydrogen to form ammonia. Mainly used to manufacture fertilizer.

habitat An *ecosystem* such as woodland or grassland that supports characteristic communities of animals and plants.

HANPP – Human appropriation of net primary production A measure of human use of the Earth's photosynthetic productivity. Net primary production is the net amount of solar energy converted to plant matter. HANPP is seen in, for example, the use of net primary production as food, wood, paper, and plant fibres.

hydro energy Energy obtained from falling or flowing water; for example, in hydroelectric power, where water is used to turn a turbine and generate electricity.

hydropower Electrical energy obtained from falling or flowing water; for example as produced by turbines in a hydroelectric dam.

ice sheet A mass of glacial land ice covering more than 50,000 km^2 (20,000 sq miles). Earth's two main ice sheets lie over Greenland and Antarctica.

infrared A form of electromagnetic energy whose waves are just longer than those of visible light. Some of the Sun's energy, and some of the heat from Earth's surface, is in the form of infrared radiation.

inundation The overflow of water to cover normally dry land, as when a river floods or a storm surge occurs on a coast.

invasive alien species A species that is not native to a particular ecosystem and causes harm when introduced to that system.

invertebrate An animal with no backbone. Such animals include insects, molluscs, crustaceans, and worms.

IUCN red list The register of animals, plants, and fungal species across the world that are deemed to be at some risk of extinction.

La Niña A large-scale change in temperature occurring about every three to seven years in the central and eastern equatorial Pacific Ocean, during which the ocean surface is cooler than normal, causing disruption to weather, especially in the Americas, Australia, and Southeast Asia. Counterpart of *El Niño*.

Latin America The countries of Central and South America, primarily those where the populations mainly speak Spanish, Portuguese, or French.

least developed country (LDC) A country with very low per capita incomes; LDCs are the poorest of the *developing countries*.

literacy The ability to read and write. Literacy, especially for women and children, is a key indicator of economic and social development in a country.

malnourishment Having a diet that does not contain the right balance of nutrients: for example, too little vitamin C or too little protein. See also *Undernourishment*.

megacity A city and its surrounding area that has more than 10 million people, such as Tokyo, New York, or São Paulo.

methane A colourless, highly flammable, gaseous hydrocarbon. Methane is the main component of natural gas and a very powerful *greenhouse gas*. Globally, more than 60 per cent of human-induced emissions arise from industry, agriculture, and landfills.

millennium development goals A set of eight development goals (including one relating to the environment) set out by the United Nations (UN) in 2000, to be achieved by 2015. Now superseded by the UN's 17 Sustainable Development Goals.

monoculture The agricultural practice of producing a single crop, plant, or livestock species, variety, or breed in a field or farming system at a time.

monsoon A seasonal change in weather, often associated with the Indian subcontinent, in which changes of wind direction and air pressure cause strong sea breezes bringing torrential summer rains.

MTOE – Million tons of oil equivalent (See *Units*).

multilateral environmental agreement (MEA) A legally binding agreement between three or more states relating to environmental issues. There are currently more than 250 MEAs in force.

MWh – Megawatt hour (See *Units*).

natural gas A fossil fuel comprised mainly of *methane*. Extracted from rocks, it is often associated with oil deposits. It is extracted by drilling or by *fracking*.

nitrous oxide A pollutant and greenhouse gas. The atmosphere naturally holds a tiny amount of nitrous oxide, but levels have markedly increased due to human activity.

nuclear power Splitting of atoms of certain elements (nuclear fission) to release energy, which is used to generate electricity. Nuclear power produces low *carbon dioxide* emissions, but the waste is highly toxic for many years.

nutrient cycle The circulation of biological and chemical matter, such as carbon or nitrogen, between the physical environment and living organisms and back again in a particular *ecosystem*.

OECD countries Countries belonging to the Organisation for Economic Cooperation and Development, a body set up by the most developed countries in 1968 to promote economic development and social progress. There are 34 OECD countries.

organic farming A method of agriculture in which farmers avoid the use of manufactured pesticides and fertilisers, instead relying on more natural processes to maintain soil fertility, including animal manure and nitrogen-fixing plants.

ozone A colourless gas that can be harmful to plants and animals in the air we breathe, but in the upper atmosphere protects the Earth from the Sun's ultraviolet radiation. Ozone concentration is measured using *Dobson units*.

ozone layer A layer of the atmosphere, 20–50 km (12–31 miles) from the Earth's surface, that contains relatively high concentrations of ozone. Thinning of the ozone layer can expose organisms (including humans) to dangerous levels of ultraviolet radiation.

permafrost Soil or rock that has remained continuously frozen for more than 2 years. In some areas, such as Alaska and Siberia, permafrost has existed for thousands of years.

persistent organic pollutants (POPs) Chemical compounds that resist being broken down and remain in the environment for a long time. Some POPs, such as DDT, are harmful to wildlife and human health.

petrochemicals Chemical compounds derived from petroleum or *natural gas*. They are used in thousands of products, such as solvents, detergents, plastics, and synthetic fibres.

photochemical smog A form of air pollution that occurs when sunlight reacts with nitrogen oxide and *volatile organic compounds*, making the air foggy or hazy. The smog can contain *ozone* and be harmful to breathe.

photosynthesis The process by which plants and some microorganisms use the energy from sunlight to convert *carbon dioxide* and water into glucose, releasing oxygen as a waste product.

photovoltaic system A technology in which photovoltaic cells or panels convert sunlight into electricity. PV systems produce clean and renewable power.

phytoplankton Tiny forms of *plankton* that live in the sunlit upper layers of oceans and lakes and use *photosynthesis* to take in *carbon dioxide* and release oxygen, thus playing a vital part in the carbon cycle.

plankton Small organisms, ranging from single-celled algae and bacteria to jellyfish, that spend part or all of their lives drifting in seas or lakes. Plankton plays vital roles in aquatic food chains. See *phytoplankton* and *zooplankton*.

polychlorinated biphenyls (PCBs) A group of man-made chemicals that were widely used in the past in products such as electrical equipment, adhesives, and paints. PCBs are *persistent organic pollutants (POPs)* that can damage health, and are now banned in many countries.

pre-industrial world The world as it was before 1750, when the Industrial Revolution began. Humans lived mainly by agriculture or small-scale industries. Pressures on the environment were much lower than they are today.

primary production The rate of conversion of solar energy into new plant *biomass* by means of *photosynthesis*.

pteropods A group of free-swimming marine snails. Pteropods have been recognised as victims of ocean *acidification*, which causes thinning of their shells.

rainforest A dense forest in a tropical or temperate area with high annual rainfall. Many rainforests are notable for *biodiversity* and are major oxygen producers and *carbon sinks*.

real GDP A measure of the value of all goods and services produced in a given year, adjusted for inflation.

recycling The conversion of domestic, agricultural, or industrial waste products into new usable materials. Recycling helps to save energy and reduce pollution.

renewable energy A term for an energy source (for power, heat, or transport) that can be constantly replenished instead of being progressively depleted. Examples include solar, wind, and hydro-power.

savanna woodland A form of tropical vegetation consisting mainly of open grassland together with scattered trees and bushes.

seabed The floor of a sea or ocean.

sulphur dioxide An air pollutant primarily emitted by burning *fossil fuels* such as coal. Sulphur dioxide can mix with water vapour to form *acid rain*; it is also a health hazard to animals and humans.

sustainability The term to describe the circumstances in which a human activity can continue indefinitely into the future, for example in relation to farming, energy generation, waste management, forestry or materials consumption.

TWh – Terawatt hour (See *Units*).

turnover The total amount that an organization earns, before taxes and other costs, from selling goods or services during a specific time period.

ultraviolet light A form of electro-magnetic energy whose waves are just shorter than those of visible light. Some of the Sun's energy is in the form of ultraviolet (UV-A and UV-B) radiation, most of which is blocked by the Earth's *atmosphere* before it reaches the surface.

undernourishment A consequence of consuming too few essential nutrients or using or excreting them more rapidly than they can be replaced. See also *Malnourishment*.

urbanization The process by which large numbers of people come together to live and work in relatively small areas, forming towns and cities.

urban density A measure of the intensity of human land use in an urbanized area, such as the number of people or the total floor area of buildings per km^2/sq mile.

UV radiation See *Ultraviolet light*.

vertebrate An animal with a backbone and an internal skeleton. Vertebrates include fish, amphibians, reptiles, birds, and mammals.

volatile organic compounds (VOCs) Carbon-based chemical compounds that evaporate readily. Found in man-made substances such as fuels, pesticides, and solvents, VOCs are air pollutants that can cause *photochemical smog*.

water table In soil or rock below the ground surface, the level below which the rock is saturated with *groundwater*.

weather The day-to-day atmospheric conditions in a particular place; aspects include air temperature and pressure, hours of sunshine, cloud cover, humidity, and rainfall or snowfall.

weathering The breakdown of rock in situ (in a specific place) at the ground surface, by wind, water, temperature changes, or chemical reactions. See also *Erosion*.

zooplankton Animals that live part or all of their life as *plankton*. They include amoebae, the larvae and juveniles of fish, and the larvae of molluscs, crustaceans and jellyfish. Zooplankton feed on *phytoplankton* and in turn are a vital food source for larger animals.

Index

Page numbers in **bold** indicate main references

A

B

C

F

G

H

I

Q

R

S

T

References and acknowledgments

Dorling Kindersley would like to thank the following:
Hugh Schermuly and Cathy Meeus for work on the original concept for this book; Peter Bull for feature illustrations; Andrea Mills, Nathan Joyce, and Martyn Page for additional editorial work; Katherine Raj and Alex Lloyd for design assistance; Katie John for proofreading and the glossary; Hilary Bird for the index; Vicky Richards for editorial research assistance; Myriam Megharbi for picture research and credits.

Main references

pp16-17: UN, Department of Economic and Social Affairs, Population Division (2013), World Population Prospects: "Most populous countries, 2014 and 2050", 2014 World Population Data Sheet, Population Reference Bureau, http://www.prb.org; Revised data: World Bank: https://data.worldbank.org/indicator/SP.POP.TOTL?locations=BR-CN-IN-ID-US; Quote from Al Gore: featured in O, The Oprah Magazine, February 2013; **pp18-19:** UN, Department of Economic and Social Affairs, Population Division (2013), World Population Prospects; "Africa will be home to 2 in 5 children by 2050: Unicef Report", Unicef press release, 12th Aug 2014, http://www.unicef.org; **pp20-21:** UN, Department of Economics and Social Affairs, Population Division. World Population Prospects, the 2015 revision; "Correlation between fertility and female education", European Environment Agency, 2010,http://www.eea.europa.eu; Latest population estimate https://www.worldometers.info/world-population/; **pp22-23**: Correlation between fertility and female education data from the European Environment Agency, 2010. Accessed at http://www.eea.europa.eu/data-and-maps/figures/correlation-between-fertility-and-female-education **pp24-25:** Global GDP https://ourworldindata.org/economic-growth; Per capita income https://ourworldindata.org/grapher/gdp-per-capita-worldbank?tab=chart®ion=World The World Economy—50 Years of Near Continuous Growth, Dariana Tani, World Economics, March 2015, http://www.worldeconomics.com; Quote by Kenneth Boulding in: United States. Congress. House (1973) Energy reorganization act of 1973: Hearings; **pp28-29:** GDP per capita, World Development Indicators, World Bank national accounts data, and OECD National Accounts data files, The World Bank, 2015, http://www.worldbank.org; 2019 data from https://data.worldbank.org/indicator/NY.GDP.PCAP.CD 2011; **pp30-31:** The World in 2015: Will the shift in global economic power continue?, PricewaterhouseCoopers LLP, February 2015; Exhibit from "Urban economic clout moves east", March 2011, McKinsey Global Institute, www.mckinsey.com/mgi. © 2011 McKinsey & Company. All rights reserved. Reprinted by permission; **pp32-33:** Exports of goods and services (current US$), World Bank national accounts data, and OECD National Accounts data files, The World Bank http://www.worldbank.org; Top U.S Trade Partners, US Department of Commerce International Trade Administration, http://www.trade.gov; **pp34-35:** World Urbanization Prospects 2014, The Department of Economic and Social Affairs of the UN Secretariat, Highlights 2014; Main graphic- World Bank: https://data.worldbank.org/indicator/SP.URB.TOTL.IN.ZS; quote by George Monbiot, published on the Guardian's website, 30th June 2011 http://www.monbiot.com/2011/06/30/atro-city/; **pp36-37:** World Urbanization Prospects 2014, The Department of Economic and Social Affairs of the UN Secretariat, Highlights 2014; http://www.un.org/en/development/desa/population/publications/pdf/urbanization/the_worlds_cities_in_2016_data_booklet.pdf; **pp38-39:** City Limits: A resource flow and ecological footprint analysis of Greater London (2002), commissioned by IWM (EB) Chartered Institute of Wastes Management Environmental Body, 12th September 2002, http://www.citylimitslondon.com; "If the world's population lived like...", Per Square Mile, Tim de Chant, August 8 2012, http://persquaremile.com; **pp40-41:** Main graphic from https://ourworldindata.org/energy; China's coal consumption from https://www.iea.org/reports/coal-information-overview;Quote by Desmond Tutu from The Guardian, September 10, 2015; **pp42-43:** Energy and Climate Change, World Energy Outlook Special Report, International Energy Agency, 2015; **pp44-45:** The Rough Guide to Green Living, Duncan Clark, Rough Guides, 2009, p26; **pp46-47:** Global renewable electricity production by region and global total via https://www.iea.org/articles/renewables-2020-data-explorer?mode=market®ion=World&product=Total "Not a toy: Plummeting prices are boosting renewables, even as subsidies fall", The Economist, April 9th 2015; **pp50–51:** Main graphic https://www.statista.com/statistics/217522/cumulative-installed-capacity-of-wind-power-worldwide/ Quote by Arnold Schwarzenegger, BBC news, April 2012; **pp56–57:** FAO http://www.fao.org/worldfoodsituation/csdb/en/#:~:text=Nonetheless%2C%20global%20cereal%20production%20is%20still%20expected%20to,cut%20by%206.8%20million%20tonnes%20month%20on%20month and World Bank https://data.worldbank.org/indicator/AG.PRD.CREL.MT?locations =CN-1W; Global Grain Production 1950–2012, Compiled by Earth Policy Institute from U.S. Department of Agriculture (USDA), http://www.earth-policy.org; Global Grain Stocks Drop Dangerously Low as 2012, Consumption Exceeded Production, J. Larson, Earth Policy Institute, January 17, 2013; World Agriculture Towards 2015/2030: An FAO Perspective, edited by J. Bruinsma, Earthscan Publications, Food and Agriculture Organization, 2003; Quote by Norman Borlaug, Nobel lecture December 11, 1970; **pp58–59:** World Bank https://data.worldbank.org/indicator/AG.PRD.CREL.MT?locations=CN-1W; The State of the World's Land and Water Resources for Food and Agriculture: Managing systems at risk, The Food and Agriculture Organization of the UN and Earthscan, 2011; The importance of three centuries of land-use change for the global and regional terrestrial carbon cycle, Climate Change, 97, 2 July 2009, pp123–144; Utilisation of World Cereal Production, Hunger in Times of Plenty, Global Agriculture, http://www.globalagriculture.org; **pp60–61:** Source: https://ourworldindata.org/fertilizer-and-pesticides#fertilizer-consumption Max Roser (2015) – 'Fertilizer and Pesticides'. Published online at OurWorldInData.org. Retrieved from: http://ourworldindata.org/data/food-agriculture/fertilizer-and-pesticides/; **pp62–63:** Active ingredients in pesticides from https://croplife.org/wp-content/uploads/2018/ 11/Phillips-McDougall-Evolution-of-the-Crop-Protection-Industry-since-1960-FINAL.pdf; "We've covered the world in pesticides. Is that a problem?", Brad Plumer, The Washington Post, Aug 18, 2013; Max Roser (2015) "Fertilizer and Pesticides" Published online at OurWorldInData.org. http://ourworldindata.org/data/food-agriculture/fertilizer-and-pesticides/ ; Popular Pesticides Linked to Drops in Bird Populations, by Helen Thompson, Smithsonian Magazine, July 2014, http://www.smithsonianmag.com/; **pp64–65:** http://www.fao.org/save-food/resources/keyfindings/infographics/fish/en/ SAVE FOOD: Global Initiative on Food Loss and Waste Reduction, Food and Agriculture Organization of the UN, http://www.fao.org; **pp66–67:** Prevalence of undernourishment data from Food and Agriculture Organization http://www.fao.org/sustainable-development-goals/indicators/2.1.1/en/; America Spends Less on Food Than Any Other Country, Alyssa Battistoni, Mother Jones, Wed Feb. 1, 2012, http://www.motherjones.com/; Quote from John F. Kennedy courtesy of the American Presidency Project; **pp68–69:** Restoring the land, Dimensions of need—An atlas of food and agriculture, FAO, Rome, Italy, 1995, http://www.fao.org; Natural Resources and Environment, FAO, 2015; **pp70–71:** "Great Acceleration", International Geosphere-Biosphere Programme, 2015, http://www.igbp.net; Trends in global water use by sector, Vital Water Graphics: An Overview of the State of the World's Fresh and Marine Waters, UN Environment Programme/GRID-Arendal, 2008, http://www.unep.org; Water withdrawal and consumption: the big gap, Vital Water Graphics: An Overview of the State of the World's Fresh and Marine Waters, UN Environment Programme/GRID-Arendal, 2008; Quote by Lyndon B Johnson, from letter to the President of the Senate and to the Speaker of the House, November 1968; **pp72–73:** Total Renewable Freshwater Supply by Country (2013 Update), http://worldwater.org; **pp76–77:** National Water Footprint Accounts: The Green, Blue, and Grey Water Footprint of Production and Consumption, M.M. Mekonnen and A.Y. Hoekstra, Value of Water Research Report Series No.50, UNESCO-IHE Institute for Water Education, May 2011; "Product Gallery", Interactive Tools, Water Footprint Network, http://waterfootprint.org; Living Planet Report 2010, Global Footprint Network, Zoological Society London, World Wildlife Fund, http://wwf.panda.org; **pp78–79:** "Addicted to resources", Global Change, International Geosphere-Biosphere Programme, April 10, 2012, http://www.igbp.net; Consumption and Consumerism, Anup Shah, January 05, 2014, http://www.globalissues.org; "Waste from Consumption and Production—Our increasing appetite for natural resources", Vital Waste Graphics, GRID-Arendal 2014, http://www.grida.no; Quote by Pope Francis, from a letter to Australia Prime Minister Tony Abbott, chair of the conference of G20 nations, November 2014; World consumes 100 billion tonnes of resources https://www.theguardian.com/environment/2020/jan/22/worlds-consumption-of-materials-hits-record-100bn-tonnes-a-year; **pp80–81:** "Bottled Water", compiled by Stefanie Kaiser, Dorothee Spuhler, Sustainable Sanitation and Water Management, http://www.sswm.info/; Growth in consumption from Statista; "New NIST Research Center Helps the Auto Industry 'Lighten Up'", Mark Bello, Centre for Automotive Lightweighting (NCAL), National Institute of Standards and Technology (NIST), August 26, 2014, http://www.nist.gov/; "Passenger Car Fleet Per Capita", European Automobile Manufacturers Association, 2015. http://www.acea.be/statistics/tag/category/passenger-car-fleet-per-capita; **pp82–83:** "When Will We Hit Peak Garbage?", Joseph Stromberg, Smithsonian Magazine, Oct 30, 2013, http://www.smithsonianmag.com; Status of Waste Management, Dennis Iyeke Igbinomwanhia, Integrated Waste Management–Volume II, edited by Sunil Kumar, August 23, 2011; "Solid Waste Composition and Characterization: MSW Materials Composition in New York State", New York State Department of Environmental Conservation, 2015, http://www.dec.ny.gov; 9 Million Tons of E-Waste Were Generated in 2012, Felix Richter, Statista, May 22, 2014, http://www.statista.com/; **pp84–85:** Municipal waste by treatment operation https://www.oecd-ilibrary.org/sites/f5670a8d-en/index.html?itemId=/content/component/f5670a8d-en; **pp86–87:** CAS Assigns the 100 Millionth CAS Registry Number to a Substance Designed to Treat Acute Myeloid Leukemia, Chemical Abstracts Service (CAS): A division of the American Chemical Society, June 29, 2015, https://www.cas.org; **pp88–89:** Quote by Sir David Attenborough from launch of World Land Trust's (WLT) first wildlife webcam, Jan 2008. http://www.worldlandtrust.org/; **pp96–97:** World internet use https://data.worldbank.org/indicator/IT.NET.USER.ZS ,connectivity: Economic and social benefits of expanding internet access, Deloitte, 2014, http://www2.deloitte.com; Quote by Kofi Annan, in opening address to the 53rd annual DPI/NGO Conference, 2006; **pp92–93:** https://www.itu.int/en/ITU-D/Statistics/Documents/facts/ICTFactsFigures2017.pdf; The Rise of Mobile Phones: 20 Years of Global Adoption", SooIn Yoon, Cartesian, June 29, 2015, http://www.cartesian.com; Smartphone usage https://leftronic.com/blog/smartphone-usage-statistics ; Social media globally https://datareportal.com/reports/digital-2020-global-digital-overview; **pp94–95:** Main graphic: https://data.worldbank.org/indicator/IS.AIR.PSGR; Top Flight Routes: http://www.iata.org/pressroom/pr/Pages/2016-07-05-01.aspx [; http://www.panynj.gov/airports/pdf-traffic/ATR2016.pdf; Air transport, passengers carried, World Development Indicators, International Civil Aviation Organization, Civil Aviation Statistics of the World and ICAO staff estimates, The World Bank, http://www.worldbank.org; "300 world 'super routes' attract 20% of all air travel", Amadeus, 16 April 2013, http://www.amadeus.com; 2019 flight figures https://www.icao.int/annual-report-2019/Pages/the-world-of-air-transport-in-2019.aspx; **pp96–97:** Source: https://data.worldbank.org/topic/poverty; Max Roser (2016)–"World Poverty". Published online at OurWorldInData.org. Retrieved from: http://ourworldindata.org/data/growth-and-distribution-of-prosperity/world-poverty; 5 Reasons Why 2013 Was The Best Year In Human History, Zack Beauchamp, ThinkProgress, Dec 11, 2013, http://thinkprogress.org; World Development Indicators 2015 maps, The World Bank, 2015, http://data.worldbank.org/maps2015; Quote by Ban Ki-moon, "Sustainable energy for all a priority for UN secretary-general's second term," New York, September 21, 2011; **pp98–99:** Proportion of population using improved drinking-water sources, Rural: 2012, WHO, 2014. http://www.who.int/en; proportion of population using improved sanitation facilities, WHO, Total: 2012, WHO, 2014; **pp100–101:** Education: Literacy rate, UNESCO Institute of Statistics, UN Educational, Scientific and Cultural Organization, 23 Nov 2015, http://data.uis.unesco.org; **pp102-103:** Maternal mortality statistics from https://data.unicef.org/topic/maternal-health/maternal-mortality/; Main graphic: http://www.who.int/healthinfo/global_burden_disease/estimates/en/index1.html; Causes of death, by WHO region, Global Health Observatory, WHO, http://www.who.int; The 10 leading causes of death by country income group (2012), Media Centre, WHO; **pp104–105:** GDP per capita (current US$), World Development Indicators, World Bank national accounts data, and OECD National Accounts data files, http://www.worldbank.org; Country Comparison: Distribution of Family Income – GINI Index, The World Factbook, Central Intelligence Agency, https://www.cia.gov; 2015 Billionaire Net Worth as Percent of Gross Domestic Product (GDP) by Nation, Areppim, 24 April 2015, http://stats.areppim.com/stats/stats_richxgdp.htm; **pp108-109:** UNHCR https://www.unnhcr.org/figures-at-a-glance.html; **pp110–11:** Main carbon dioxide graphic www.esrl.noass.gov/gmd/ccgg/trends/ Methane and nitrous oxide from same source; Quote from Leonardo di Caprio: address to UN Climate Summit, New York, Sept 2014 pp120-21: Sources of greenhouse gases detailed in 'IPCC, 2014: Summary for Policymakers', Contribution of Working Group III to the Fifth Assessment Report of the Intergovernmental Panel on Climate Change. Available at http://www.ipcc.ch/pdf/assessment-report/ar5/wg3/

ipcc_wg3_ar5_summary-for-policymakers.pdf pp122-23: Ozone depletion maps based on images compiled by NASA. Accessed at http://earthobservatory.nasa.gov/Features/WorldOfChange/ozone.php?all=y 2020 ozone hole source from https://svs.gsfc.nasa.gov/4882 pp124-25: Main temperature rise graphic based on data from 'IPCC Fifth Assessment Report - Climate Change 2013: The Physical Science Basis', Intergovernmental Panel on Climate Change (IPCC), 2013. Accessed at https://www.ipcc.ch/pdf/assessment-report/ar5/wg1/WGIAR5_SPM_brochure_en.pdf "Rising waters" graphic uses data from 'Impact of sea level rise in Bangladesh', United Nations Environment Programme, 2008. Accessed at http://www.unep.org/dewa/vitalwater/article146.html; **pp112-13:** Sources of greenhouse gases detailed in 'IPCC, 2014: Summary for Policymakers', Contribution of Working Group III to the Fifth Assessment Report of the Intergovernmental Panel on Climate Change. Available at http://www.ipcc.ch/pdf/assessment-report/ar5/wg3/ipcc_wg3_ar5_summary-for-policymakers.pdf; **pp114-15** Ozone depletion maps based on images compiled by NASA. Accessed at http://earthobservatory.nasa.gov/Features/WorldOfChange/ozone.php?all=y; **pp116-17:** Main temperature rise graphic based on data from 'IPCC Fifth Assessment Report - Climate Change 2013: The Physical Science Basis', Intergovernmental Panel on Climate Change (IPCC), 2013. Accessed at https://www.ipcc.ch/pdf/assessment-report/ar5/wg1/WGIAR5_SPM_brochure_en.pdf' "Rising waters" graphic uses data from 'Impact of sea level rise in Bangladesh', United Nations Environment Programme, 2008. Accessed at http://www.unep.org/dewa/vitalwater/article146.html. the map of seasonal ice melt in the Arctic uses information from 'The Future of Arctic Shipping', Malte Humpert and Andreas Raspotnik,
The Arctic Institute, 2012. Accessed at http://www.thearcticinstitute.org/the-future-of-arctic-shipping/; **pp120-21:** Summer flounder stirs north-south climate change battle, Marianne Lavelle, The Daily Climate, June 3, 2014; Top scientists agree climate has changed for good, Sarah Clarke, ABC news, 3 April 2013, http://www.abc.net.au; Spring is Coming Earlier, Climate Central, Mar 18th, 2015, http://www.climatecentral.org; **pp124-25:** Data for 1.5 degrees from IPCC https://www.ipcc.ch/sr15/chapter/spm/; **pp126-27:** The 2010 Amazon Drought, Science, 04 Feb 2011, Vol.331, Issue 6017, pp554, http://science.sciencemag.org; **pp128-29:** The Unburnable Carbon Concept Data 2013, Carbon Tracker Initiative, September 17, 2014, http://www.carbontracker.org; carbon budget for 1.5 degrees https://www.ipcc.ch/sr15/chapter/spm/; China carbon emissions 28 per cent of global total https://www.bbc.co.uk/news/science-environment-54256826; **pp130-31:** IPCC, 2014: Climate Change 2014: Synthesis Report. Contribution of Working Groups I, II and III to the 5th Assessment Report of the Intergovernmental Panel on Climate Change; Quote from Pope Francis, at meeting with political, business and community leaders, Quito, Ecuador, July 7, 2015; **pp132-33:** "Deforestation Estimates: Macro-scale deforestation estimates (FAO 2010)," Monga Bay, http://www.mongabay.com; **pp134-35:** "6 Graphs Explain the World's Top 10 Emitters", Mengpin Ge, Johannes Friedrich and Thomas Damassa, World Resources Institute, November 25, 2014; Quote from Barack Obama, taken from speech at the GLACIER Conference, Anchorage, Alaska, 1 September, 2015; **pp136-37:** "Desolation of smog: Tackling China's air quality crisis", David Shukman, BBC News: Science and Environment, 7 January 2014, http://www.bbc.co.uk; Burden of disease from Ambient Air Pollution for 2012, WHO, 2014, http://www.who.int; **pp140-141:** Global human appropriation of net primary production doubled in the 20th century, Proceedings of the National Academy of Sciences of the United States of America, 2013, http://www.pnas.org; "Of Fossil Fuels and Human Destiny," Peak Oil Barrel, http://peakoilbarrel.com; Quote from the HRH The Prince of Wales from Presidential Lecture, Presidential Palace, Jakarta, Indonesia, November 2008; **pp142-143:** Forest Resources Assessment, Food and Agriculture Organization of the UN, 2020 http://www.fao.org/forest-resources-assessment/2020/en/;
pp144-45: Lake Chad - decrease in area 1963, 1973, 1987, 1997 and 2001, Philippe Rekacewiz, UNEP/GRID-Arendal 2005 , http://www.grida.no; **pp146-47:** IFPRI (International Food Policy Research Institute). 2012. "Land Rush" map. Insights 2 (3). Washington, DC: International Food Policy Research Institute. http://insights.ifpri.info/2012/10/land-rush/; **148-49:** State of the world's Fisheries and Aquaculture, Food and Agriculture Organization of the UN, 2020, : http://www.fao.org/state-of-fisheries-aquaculture; Collapse of Atlantic cod stocks off the East Coast of Newfoundland in 1992, Millennium Ecosystem Assessment, 2007, Philippe Rekacewicz, Emmanuelle Bournay, UNEP-GRID-Arendal, http://www.grida.no; Good Fish Guide, Marine Conservation Society, 2015, http://www.fishonline.org; Quote from Ted Danson, reported in New York Times, "What's worse than an oil spill?", April 20, 2011; **pp150-51:** Good Fish Guide, Marine Conservation Society, 2015, http://www.fishonline.org; **pp154-55:** "Top Sources of Nutrient Pollution" and "The

Eutrophication Process," Ocean Health Index 2015, http://www.oceanhealthindex.org; N.N. Rabalais, Louisiana Universities Marine Consortium and R.E. Turner, Louisiana State University, http://www.noaanews.noaa.gov/stories2013/2013029_deadzone.html; **pp156–57:** 22 Facts About Plastic Pollution (And 10 Things We Can Do About It), Lynn Hasselberger, The Green Divas, EcoWatch, April 7, 2014, http://ecowatch.com; "When The Mermaids Cry: The Great Plastic Tide", Claire Le Guern Lytle, Plastic Pollution, Coastal Care, http://plastic-pollution.org; **pp158–59:** GLOBIO3: A Framework to Investigate Options for Reducing Global Terrestrial Biodiversity Loss, Ecosystems (2009), 12, pp374–390, Rob Alkenmade, Mark van Oorschot, Lera Miles, Christian Nellemann, Michel Bakkenes, and Ben ten Brink, http://www.globio.info; Accelerated modern human–induced species losses: Entering the sixth mass extinction, Gerardo Ceballos, Paul R. Ehrlich, Anthony D. Barnosky, Andrés García, Robert M. Pringle and Todd M. Palmer, Science Advances, 19 June 2015, http://advances.sciencemag.org; Defaunation in the Anthropocene, Science, 25 July 2014, Vol. 345. Ossie 6195, pp401-406, http://science.sciencemag.org; Quote from Sir David Attenborough during Q&A session on social media site Reddit, 8 January 2014;

pp160–61: "Where we work", Critical Ecosystem Partnership Fund, http://www.cepf.net; **pp168–69:** Changes in the global value of ecosystem services, Robert Costanza et al, Global Environmental Change, 26, Elsevier, 1 April 201; Quote by Satish Kumar, reported in Resurgence and Ecologist, 29th August 2008; **pp170–71:** 'This pandemic is an environmental issue' https://www.tonyjuniper.com/content/pandemic-environmental-issue; **pp172–73:** Data on COVID-19 deaths, John Hopkins University https://coronavirus.jhu.edu/data/mortality Economic impact of pandemic, University of Cambridge https://www.jbs.cam.ac.uk/insight/2020/economic-impact/ Impact on carbon emissions, Carbon Brief https://www.carbonbrief.org/daily-brief/global-emissions-plunged-an-unprecedented-17-percent-during-the-coronavirus-pandemic; **pp174–75:** Quote from Sir Jonathon Porritt, in "Capitalism as if the world matters", first published 2005; **pp176–77:** "The Age of Humans: Evolutionary Perspectives on the Anthropocene", Human Evolution Research, Smithsonian National Museum of Natural History, 16 November 2015; "The Anthropocene is functionally and stratigraphically distinct from the Holocene", Science, Vol. 351, Issue 6269, http://science.sciencemag.org; Quote by Will Steffen from report of the IGBP, January 2015; **pp178–79:** "The Nine Planetary Boundaries", 2015, Stockholm Resilience Centre Sustainability Science for Biosphere Stewardship, http://www.stockholmresilience.org; "How many Chinas does it take to support China?", Infographics, Earth Overshoot Day 2015, http://www.overshootday.org; **pp180–81:** Water Consumption for Operational Use by Energy Type, Climate Reality Project, October 05 2015, https://www.climaterealityproject.org; **pp182–83:** Ratification of multilateral environmental agreements, Riccardo Pravettoni, UNEP/GRID-Arendal, http://www.grida.no; 100 Years of Multilateral Environmental Agreements, Plotly, 2015, https://plot.ly/~caluchko/39/_100-years-of-multilateral-environmental-agreements; **pp184–85:** "Sustainable Development Goals: 17 Goals to Transform Our World", UN, 2015, http://www.un.org; Global Biodiversity Outlook 5 https://www.cbd.int/gbo/gbo5/publication/gbo-5-spm-en.pdf; **pp186–87:** Measuring Progress: Environmental Goals & Gaps, UN Environment Programme (UNEP), 2012, Nairobi, http://www.unep.org; The Millennium Development Goals Report 2015, UN, New York, 2015, http://www.un.org; **pp188–89:** Global Biodiversity Outlook 5 https://www.cbd.int/gbo/gbo5/publication/gbo-5-spm-en.pdf; **pp190–91:** Deguignet M., Juffe-Bignoli D., Harrison J., MacSharry B., Burgess N., Kingston N., (2014) 2014 UN List of Protected Areas, UNEP-WCMC: Cambridge, UK, http://www.unep-wcmc.org; **pp192–93:** Figure 2, "Waves of Innovation of the First Industrial Revolution", TNEP International Keynote Speaker Tours, The Natural Edge Project, 2003-2011, http://www.naturaledgeproject.net; "Biomimicry Examples", The Biomimicry Institute, 2015, http://biomimicry.org; **pp194–95:** Prosperity without Growth?, The Sustainable Development Commission, Professor Tim Jackson, March 2009, http://www.sd-commission.org.uk; Two degrees of separation: ambition and reality. Low Carbon Economy Index 2014, PricewaterhouseCoopers LLP, September 2014, http://www.pwc.co.uk; **pp196–197:** Clean green jobs data from https://www.irena.org/-/media/Files/IRENA/Agency/Publication/2020/Sep/IRENA_RE_Jobs_2020.pdf "Small and Medium-sized Enterprises can Unlock $1.6 trillion Clean Tech Market in next 10 years", The Climate Group, 25 September 2014, http://www.theclimategroup.org; infoDev. 2014. Building Competitive Green Industries: The Climate and Clean Technology Opportunity for Developing Countries. Washington, DC: World Bank. License: Creative Commons Attribution CC BY 3.0, http://www.infodev.org; IRENA (2014), Renewable Energy and Jobs–Annual Review 2014, International Renewable Energy Agency, http://www.irena.org; **pp198–99:** Rewiring the Economy,

Cambridge Institute for Sustainability Leadership, 2015, http://www.cisl.cam.ac.uk; **pp202–203:** "Circular Economy", The Ellen MacArthur Foundation, http://www.ellenmacarthurfoundation.org; "Phosphorus Recycling", Friends of the Earth Sheffield, 2013, http://planetfriendlysolutions.blogspot.co.uk; **pp200–201:** "A Safe and Just Space for Humanity: Can we live within the doughnut?", Kate Raworth, Oxfam Discussion Papers, Oxfam International, February 2012, https://www.oxfam.org; **pp204–205:** quote from Ban Ki-moon, Remarks to the General Assembly on his Five-Year Action Agenda: "The Future We Want" 25 January, 2012.

Data citations

24-25 Data from: Max Roser (2013) - "Economic Growth". Published online at OurWorldInData.org. Retrieved from: 'https://ourworldindata.org/economic-growth' [Online Resource], CC BY 4.0. **28-29** Data from: Max Roser (2013) - "Economic Growth". Published online at OurWorldInData.org. Retrieved from: 'https://ourworldindata.org/economic-growth' [Online Resource], CC BY 4.0. **40-41** Data from: https://ourworldindata.org/energy-production-consumption. **41 IEA:** Data from: IEA (2020), Coal Information: Overview, IEA, Paris https://www.iea.org/reports/coal-information-overview. All rights reserved (as modified by DK). **47 IEA:** Data from: IEA (2020), Renewables 2020 Data Explorer, IEA, Paris https://www.iea.org/articles/renewables-2020-data-explorer. All rights reserved (as modified by DK) (tr). **50 Statista:** Data from: "Cumulative installed capacity of wind power worldwide in 2019, by country" - https://www.statista.com/statistics/217522/cumulative-installed-capacity-of-wind-power-worldwide/ (r). **56-57 Food and Agriculture Organization of the United Nations:** Data from: © FAO 2021, "World cereal markets heading towards a record production in 2021/22 but only a marginal increase foreseen for stocks" - http://www.fao.org/worldfoodsituation. **67 Food and Agriculture Organization of the United Nations:** Data from: © FAO, "Percentage of undernourished people by region in 2000 and 2019" - http://www.fao.org/sustainable-development-goals/indicators/2.1.1/en/ - accessed 2021. **70-71** Data from: Hannah Ritchie and Max Roser (2017) - "Water Use and Stress". Published online at OurWorldInData.org. Retrieved from: https://ourworldindata.org/water-use-stress [Online Resource]. **78-79** Data from F. Krausmann et al., Growth in global materials use, GDP and population during the 20th century, Ecological Economics, Vol. 68, Issue 10, 2009, pp. 2696-2705, ISSN 0921-8009, https://doi.org/10.1016/j.ecolecon.2009.05.007. **80** Data from: The Business Research Company (bc). **Statista:** Data from: "Bottled water consumption worldwide from 2007 to 2017" - https://www.statista.com/statistics/387255/global-bottled-water-consumption/. **83** Data from the E-Waste Monitor, United Nations University via OECD (r). **84-85** Data from: OECD (2021), "Circular economy - waste and materials", in Environment at a Glance Indicators, OECD Publishing, Paris, https://doi.org/10.1787/f5670a8d-en. **90-91 World Bank:** Data from: "Individuals using the Internet (% of population). Source: International Telecommunication Union (ITU) World Telecommunication/ICT Indicators Database, CC BY-4.0 (Global internet usage). **93** Data from: "A summary of global social media users around the world" - https://www.smartinsights.com/social-media-marketing/social-media-strategy/new-global-social-media-research/ (b). **96-97 World Bank:** Data from: "Poverty headcount ratio at $1.90 a day (2011 PPP) (% of population)", World Bank, Development Research Group, CC BY-4.0. **108** Data from UNHCR Global Trends 2014 (bc). **110-111 NOAA:** Data from: "Trends in Atmospheric Carbon Dioxide", Dr. Pieter Tans, NOAA/GML (gml.noaa.gov/ccgg/trends/) and Dr. Ralph Keeling, Scripps Institution of Oceanography (scrippsco2.ucsd.edu/). **114-115 NASA's Earth Observatory:** (ozone). **143 Food and Agriculture Organization of the United Nations:** Data from: FAO. 2020. Global Forest Resources Assessment 2020 – Key findings. Rome. https://doi.org/10.4060/ca8753en. **148-149 Food and Agriculture Organization of the United Nations:** Data from: FAO. 2020. The State of World Fisheries and Aquaculture 2020. Sustainability in action. Rome.; https://doi.org/10.4060/ca9229en. **186 Carbon Brief:** Adapted from "Change in emissions in major economies since 1990", source: Climate Watch - https://www.carbonbrief.org/analysis-which-countries-met-the-uns-2020-deadline-to-raise-climate-ambition (bc/graph). **187** Data from: Climate Action Tracker (2020). Global Map. Update November 2020. Available at: https://climateactiontracker.org/countries/. Copyright © 2021 by Climate Analytics and New Climate Institute. All rights reserved. **188-189** Data from: Secretariat of the Convention on Biological Diversity (2020) Global Biodiversity Outlook 5 – Summary for; Policy Makers. Montréal. **197** Data from: IRENA (2020), Renewable Energy and Jobs – Annual Review 2020, International Renewable; Energy Agency, Abu Dhabi (Source: IRENA jobs database) (tr).

Acknowledgments

From the author

I am grateful to the many people who made this book project possible. Peter Kindersley first came up with the idea of gathering together in one place the great breadth of information that explains the profound changes taking place on Planet Earth. He provided the resources necessary to produce a proposal, during the process of which I was pleased to work with Hugh Schermuly and Cathy Meeus, who among other things provided expert assistance in producing top quality graphics. When that initial phase was completed I was very pleased to be asked to take the lead in researching and writing the title before you now. My agent Caroline Michel at Peters Fraser and Dunlop spoke with colleagues at Dorling Kindersley and made arrangements with Publishing Director Jonathan Metcalf and his team, including Liz Wheeler, Janet Mohun, and Kaiya Shang, to produce the book. Jonathan and his colleagues at Dorling Kindersley also further developed the original concept idea and ran the complex process of producing excellent graphics to convey the wealth of data we sourced. It was a pleasure to work with the design and editorial team that included Duncan Turner, Clare Joyce, Ruth O'Rourke and Jamie Ambrose.

I much appreciated early stage input to the contents from my friends and colleagues at The Prince of Wales's International Sustainability Unit (ISU), who during recent years inspired me to develop many of the ideas expressed in the book. I would like to especially mention Edward Davey who was kind enough to read and comment upon an initial overview of the title. Michael Whitehead and Claire Bradbury in the Prince of Wales's office were most helpful in facilitating the provision of the excellent foreword penned by His Royal Highness, whose efforts in taking the time to write such an excellent introduction are warmly appreciated.

My colleagues at the University of Cambridge Institute for Sustainability Leadership (CISL) provided much inspiration and insight over the years as to the nature of the trends covered in this book and I'd like to express my appreciation to them for that, including their recent work on "Rewiring the Economy". with which I was pleased to have modest involvement. I would like to express warm thanks to Madeleine Juniper for much hard work on sourcing and processing data and drafting text.

Professor Neil Burgess, Head of Science at UNEP-WCMC in Cambridge, provided a great deal of valuable advice in relation to data sources and was also kind enough to read through and comment upon an advanced draft. Rishi Modha advised on data sources relating to digital globalization, Philip Lymbery on food and farming and Jordan Walsh provided more general research assistance.

Owen Gaffney, formerly of the International Biosphere and Geosphere Programme (IGBP) in Stockholm and now with the Stockholm Resilience Centre in Sweden, provided helpful input during concept development and assisted with advice on data sources. I am indebted to Will Steffen, also at the Stockholm Resilience Centre, for inspiration regarding the concept of the Great Acceleration and for taking the time to comment on some of the draft pages.

Dr Emily Shuckburgh OBE of the British Antarctic Survey kindly provided an expert review and advice relating to the climate change and atmosphere sections of the book and for that I am very grateful.

Finally, I'd like to express appreciation and admiration to the thousands of scientists, researchers, data gatherers and number crunchers whose work enables us to know what is really happening to our planet. They work in organisations ranging from the World Bank to Oxfam and from UN specialist agencies to conservation groups. Without their efforts it would not be possible to produce such a book. Neither would it be possible without the support of my wife Sue Sparkes. We have been careful to avoid errors but if any have sneaked past the editing process then I take responsibility for that.

Dr Tony Juniper, Cambridge, January 2021

Credits

The publisher would like to thank the following for their kind permission to reproduce their photographs:

(Key: a-above; b-below/bottom; c-centre; f-far; l-left; r-right; t-top)

22 Dreamstime.com: Digitalpress (bc). **29 Getty Images:** Frederic J. Brown / AFP (br). **30 Exhibit from "Urban economic clout moves east", March 2011, McKinsey Global Institute, www.mckinsey.com/mgi. Copyright © 2011 McKinsey & Company. All rights reserved.** Reprinted by permission (b). **38 Tim De Chant:** (bl). **50 123RF.com:** tebnad (bl). **63 Dreamstime.com:** Comzeal (tr). **73 Dreamstime.com:** Phillip Gray (br). **85 123RF.com:** jaggat (tr). **92 Getty Images:** Joseph Van Os / The Image Bank (cra). **99 Dreamstime.com:** Aji Jayachandran - Ajijchan (ca). **100 Corbis:** Liba Taylor (b). **102 Dreamstime.com:** Sjors737 (bl). **108 123RF.com:** hikrcn (cb). **116 Corbis:** Dinodia (tr). **117 The Arctic Institute:** Andreas Raspotnik and Malte Humpert (br). **118 Climate Central:** http://www.climatecentral.org/gallery/maps/spring-is-coming-earlier (br). **122 123RF.com:** Meghan Pusey Diaz - playalife2006 (bl). **146 IFPRI (International Food Policy Research Institute). 2012:** "Land Rush" map. Insights 2 (3). Washington, DC: International Food Policy Research Institute. http://insights.ifpri.info/2012/10/land-rush/. Reproduced with permission. **154 Data source: N.N. Rabalais, Louisiana Universities Marine Consortium and R.E. Turner, Louisiana State University:** (bl). **161 Dreamstime.com:** Eric Gevaert (tr). **167 Dreamstime.com:** Viesturs Kalvans (bc). **172 Getty Images**: Miguel Medina / AFP (bl). **178 Source: Global Footprint Network, www.footprintnetwork.org:** (bl). **189 Dreamstime.com:** Alexandrebes (tc). **191 123RF.com:** snehit (crb). **192-193 The Natural Edge Project. 197 Getty Images:** Yang Yanmin / China News Service (crb).